World Heritage and Climate Change

World Heritage and Climate Change

Impacts and Adaptation

Editors

Chiara Bertolin
Jim Perry

MDPI • Basel • Beijing • Wuhan • Barcelona • Belgrade • Manchester • Tokyo • Cluj • Tianjin

Editors

Chiara Bertolin
Department of Mechanical and
Industrial Engineering,
Norwegian University of Science
and Technology (NTNU)
Norway

Jim Perry
Department of Fisheries,
Wildlife and Conservation Biology,
University of Minnesota System
USA

Editorial Office
MDPI
St. Alban-Anlage 66
4052 Basel, Switzerland

This is a reprint of articles from the Special Issue published online in the open access journal *Climate* (ISSN 2225-1154) (available at: https://www.mdpi.com/journal/climate/special_issues/world_heritage).

For citation purposes, cite each article independently as indicated on the article page online and as indicated below:

LastName, A.A.; LastName, B.B.; LastName, C.C. Article Title. *Journal Name* **Year**, *Volume Number*, Page Range.

ISBN 978-3-03943-943-0 (Hbk)
ISBN 978-3-03943-944-7 (PDF)

Cover image courtesy of Chiara Bertolin.

Contents

About the Editors

Chiara Bertolin is Associate Professor in "Non-destructive techniques and structural health monitoring" at the Norwegian University of Science and Technology NTNU, Department of Mechanical and Industrial Engineering. She has a master's degree and PhD in Astronomy from Padua University in Italy.

Throughout her academic career, she has developed interdisciplinary competencies, research and teaching skills on historic climatology, microclimate studies for cultural heritage preservation, analysis of climate change impacts on the built environment and the assessment of mechanical decay caused by the effects of climate variation on wood. She was awarded both the Outstanding Academic Fellow (2020-2023) and the Onsager Fellow position (2016-2021) in the Research Excellence Program at NTNU.

She is the Principal Investigator and coordinator of the he FRIPRO FRINATEK Young Research Talent International Research Project: SyMBoL—Sustainable Management of Heritage Buildings in a Long-term Perspective (2018-2021), which also involves the Norwegian Institute for Cultural Heritage Research (NIKU), the Polish Academy of Science (PAS) and the Getty Conservation Institute. Bertolin is the Leader for Norway of the ANATOLIA Project (2020–2024) for the European Space Agency (ESA) in the framework of atmospheric monitoring to assess the variability of optical links through the atmosphere. In 2020, in Nature, she published within a major international consortium of researchers, the reconstruction of the flooding characteristics in Europe over the last 500 years (Gunter Bloschl et al., 2020, Current Flood-rich period exceptional compared to past 500 years in Europe. Nature 583:560-566). In 2019, she was invited to fill the role of Contributing Author in the IPCC Working Group I Sixth Assessment Report, which is expected in 2021. Since 2011, she has been a member of the Scientific Committee of the National Standardization Body "Environment" of the Cultural Heritage Commission and of the European Committee for Standardisation CEN/TC 346—WG7 "Specifying and measuring Indoor/Outdoor Climate and Cultural Deposits". She was a Scientific Advisor for UNESCO in the Private Committees Program for the Safeguarding of Venice, for the Venice Civic Museums Foundation (MUVE) and for the Diocesan Museum in Udine, Italy.

Jim Perry is HT Morse Distinguished University Professor at the University of Minnesota. He serves as Director of Undergraduate Studies in Fisheries and Wildlife and Director of Wildlife Care and Handling. His teaching and research advance ecosystem management at the scale of large watersheds, with explicit attention to climate-based adaptation. This work focuses on resilience to advance climate change adaptation in large protected areas, notably natural World Heritage sites. His work is global and broadly applicable to watersheds as ecosystems and, more notably, to protected areas, including all natural World Heritage sites. Although the work is global in theme, it is always applied at the local scale. A recent edited volume (Harvey and Perry, 2015) reframes the ways that we consider heritage concepts as climates change. In a more focused review paper (Perry, 2015), he argues that climate change adaptation in World Heritage sites is a wicked problem (meeting several criteria for this definition) and that so-called clumsy solutions provide a way forward. Because this is a global problem, society must choose among sites to guide resource allocation. In support of such prioritization, he analyzed 208 natural UNESCO World Heritage Sites to build a global "hot spots" model that ranks sites and identifies those most at risk from climate change (Perry 2011). However, because climate change adaptation is always a local-scale action, he continued this work in collaboration with UNESCO to develop a climate change adaptation manual for managers of natural World Heritage sites or other protected areas (Perry and Falzon 2014). This manual guides any local manager in understanding the risks that climate change poses to his/her site and guides him/her toward adaptation strategies. This work was initially field-tested in Kenya and India, translated and available in three languages, and is being used worldwide. Some of the adaptation strategies offered are fine-scale (i.e., on-site) and some are coarse-scale (i.e., involving the surrounding landscape). To advance the latter, he worked with many others to develop an ecosystem-based approach to managing a watershed, catchment or landscape (e.g., one containing a World Heritage site or protected area) (Perry et al. 2012).

This ecosystem-scale work was developed in collaboration with UNEP and concluded with a global training program for ecosystem management (Perry et al., 2012). This work was initially field-tested in Kenya and then deployed in a Train-the-Trainers phase, beginning with a 12-country workshop in South Korea. All of this work has been performed in the context of greater societal goals, goals that advance society's ability to recognize and adapt to new climate regimes. This work has recently been advanced with special attention to vulnerable communities in watersheds facing climate changes (Perry et al., 2018).

Preface to "World Heritage and Climate Change"

World Heritage represents natural and cultural resources that are so exceptional that they should be forever conserved for all humankind. World Heritage sites and immovable and movable cultural heritage are of high significance, as they are a vital expression of the culture that makes a place or a community unique, and their loss could be catastrophic. However, many natural and cultural World Heritage sites are at continuous risk from climate change. There is an urgent need to understand the ways in which the climate threatens various sites, immovable objects and artifacts, and to understand what adaptation strategies, if any, are appropriate for their conservation. The literature over the last 10 years is replete with discussions of risk assessments, planning strategies and adaptation plans. We are awash with information but impeded by a dearth of understanding. This Special Issue brings together a select group of authors, each of whom contributed to an understanding of World Heritage in a changing climate. Each invited paper addresses a subset of the natural or built environment and synthesizes what is known and what uncertainties face decision-makers and offers examples of tools or strategies to implement in situ. Among the proposed strategies are novel schemes of both the mitigation of and adaptation to climate change and ideas for monitoring the conservation status of the heritage site and/or object under examination.

Detailed past climate reconstructions of dangerous events and the simulation of future climate change scenarios are lacking at the World Heritage site scale. This knowledge is important for evaluating climate changes that currently threaten World Heritages and, even more so, changes that will threaten them in the future. Scarce knowledge is available about past climate at World Heritage sites that have been reconstructed using instrumental, documentary and paleoclimatic data. The reconstruction of multi-century series of temperature, precipitation, wind force and direction, relative humidity and the frequencies of floods and landslides that impact these sites could elucidate the real effects of the changes that we are observing. This type of knowledge allows us to improve the calibration of scenarios of changes over the near and distant future and clarifies the urgency of implementing adaptation measures.

Chiara Bertolin, Jim Perry
Editors

Article

Mitigating Climate Change in the Cultural Built Heritage Sector

Elena Sesana [1], Chiara Bertolin [2], Alexandre S. Gagnon [3] and John J. Hughes [1,*]

[1] School of Computing, Engineering and Physical Sciences, University of the West of Scotland, High St, Paisley PA1 2BE, UK

[2] Department of Architecture and Technology, Norwegian University of Science and Technology (NTNU), Alfred Getz vei 3, 7491 Trondheim, Norway

[3] School of Natural Sciences and Psychology, Liverpool John Moores University, James Parsons Building, Byrom Street, Liverpool L3 3AF, UK

[*] Correspondence: john.hughes@uws.ac.uk

Received: 14 May 2019; Accepted: 8 July 2019; Published: 11 July 2019

Abstract: Climate change mitigation targets have put pressure to reduce the carbon footprint of cultural heritage buildings. Commonly adopted measures to decrease the greenhouse gas (GHG) emissions of historical buildings are targeted at improving their energy efficiency through insulating the building envelope, and upgrading their heating, cooling and lighting systems. However, there are complex issues that arise when mitigating climate change in the cultural built heritage sector. For instance, preserving the authenticity of heritage buildings, maintaining their traditional passive behaviours, and choosing adaptive solutions compatible with the characteristics of heritage materials to avoid an acceleration of decay processes. It is thus important to understand what the enablers, or the barriers, are to reduce the carbon footprint of cultural heritage buildings to meet climate change mitigation targets. This paper investigates how climate change mitigation is considered in the management and preservation of the built heritage through semi-structured interviews with cultural heritage experts from the UK, Italy and Norway. Best-practice approaches for the refurbishment of historical buildings with the aim of decreasing their energy consumption are presented, as perceived by the interviewees, as well as the identification of the enablers and barriers in mitigating climate change in the cultural built heritage sector. The findings emphasise that adapting the cultural built heritage to reduce GHG emissions is challenging, but possible if strong and concerted action involving research and government can be undertaken to overcome the barriers identified in this paper.

Keywords: adaptation; climate change; cultural heritage; energy efficiency; historical buildings; mitigation; refurbishment; sustainability

---

1. Introduction

By ratifying the Kyoto Protocol, the European Union (EU) committed to reducing its emissions of greenhouse gases (GHGs). Such a commitment was further reinforced by the 2015 Paris Agreement in which the EU agreed to a 40% reduction in GHG emissions by 2030 [1]. This can be accomplished, in part, by improving the energy efficiency of buildings as they are responsible for 40% of the total energy consumption in Europe [2,3]. However, a large percentage of the European building stock is composed of historical buildings, with 35% of them over 50 years old and 75% inefficient in their use of energy [2,4]. The potential contribution of the cultural built heritage sector to GHG emission reduction targets is therefore significant and requires action.

The United Nations Educational, Scientific and Cultural Organization (UNESCO), the International Centre for the Study of the Preservation and Restoration of Cultural Property (ICCROM) and the International Council on Monuments and Sites (ICOMOS) have pledged for the development of mitigation strategies applied to cultural heritage. This is particularly relevant to Europe where a large percentage of the heritage properties inscribed in the UNESCO World Heritage List are located [5]. They also encourage managers to reduce GHG emissions at site level, especially within the World Heritage network [6]. Strategies to mitigate climate change in the built heritage sector include the implementation of energy efficiency measures, for instance, decreasing the use of energy for lighting, heating, cooling and ventilation, or reducing the energy used for transporting building materials. Other measures which can contribute to mitigation efforts include waste reduction, reusing and recycling materials, using sustainable materials and processes and decreasing water use [7,8]. The carbon footprint of the cultural built heritage can also be reduced through energy efficiency planning and interventions to decrease emissions within its own management [9].

The cultural built heritage sector can therefore significantly contribute to climate change mitigation. The aim of this paper is to understand how climate change mitigation is currently considered in the management and preservation of the cultural built heritage in Europe. Specifically, the objectives are to determine the perspectives of experts in cultural heritage preservation on the enablers - and the barriers - to mitigate climate change within their sector, and to identify best-practice approaches for the refurbishment of historical buildings with the aim of decreasing their energy consumption.

2. Mitigating Climate Change in the Cultural Built Heritage Sector

There is an increasing body of research on climate change mitigation in the built heritage sector, with most studies focusing on reducing the energy used in heritage buildings through retrofitting efforts, i.e., improvements in the thermal performance of the building envelope, and upgrading the heating, ventilation and air conditioning systems. Less research has been accomplished on the use of traditional passive measures in historical buildings as strategies to reduce energy consumption, and on the use of the Life Cycle Assessment (LCA) methodology for the selection of materials requiring less energy to produce, and thus emitting less CO_2. There is also limited research on energy saving measures induced by changes in user behaviours, and on the challenges associated with improving the energy efficiency of heritage buildings in relation to the impact of the refurbishment on their historical value.

Several studies investigated possible retrofitting solutions for historical buildings to reduce their energy consumption. The proposed measures include improved thermal insulation of floors and roofs, external wall insulation through the use of highly insulating plaster [10], the installation of more efficient (and draught-proofing) windows, the improvement of heating, cooling, ventilation and lighting systems (e.g. installation of light emitting diodes [LEDs]), the installation of photovoltaic tiles, and even the elimination of rising damp [11–13]. Hence, previous studies have mainly presented examples of retrofitting of historical buildings that have been successful in decreasing the building energy consumption.

Improvements in the insulation of the building envelope is a major theme in climate change mitigation. There are several options and evidence for their varying effectiveness, including recently advanced options. For example, Berardi [14] investigated the properties of aerogel systems with plasters, concrete tiles/panels and fibre blankets, emphasizing that these materials have great thermal performance, but they are too expensive to be used for a sustainable economic return. Also, Zhou et al. [15] investigated the performance of internally insulated walls with aerogel-based high insulating plaster and renders such as lime mortar and mineral plaster, indicating that internal retrofitting using such materials can alter the hygrothermal performance of walls and, for this reason, recommended caution in their use. Novel but more traditional methods of insulation can also be used, e.g., Nardi et al. [16] investigated the upgrading of the internal vertical envelope using insulating panels made of hemp fibre, which resulted in increased thermal performance. There are, nonetheless, possible disadvantages resulting from these building alterations, such as an increase in decay caused by changes in hygrothermal performance and vapour movement when new materials are introduced to increase thermal performance.

Another area of attention is the improvement of the heating, cooling, ventilation and lighting systems e.g., [17], and the cogeneration of heat and power using renewable energy sources to reduce the buildings' operational energy requirements. These can include heat pumps as heating and cooling systems, which use outdoor air, underground water e.g., [18,19] and heat stored in the ground e.g., [20–22], demand-controlled ventilation and trigeneration technologies [18], efficient lighting systems [23] and 'hybrid' energy systems [24]. Unlike other research, Lo Basso et al. [24] considered the heritage values as a key factor to take into consideration when proposing changes to the heating and power system of historical buildings, notably in the use of photovoltaics, solar hybrid collectors, and heat pumps as solutions to effectively reduce their energy consumption. This is because there is a risk of incompatibility, as the use photovoltaic systems, for example, can affect the aesthetic value of a historical building. These examples, amongst others, paint a complex picture; it can be difficult to identify a solution due to the subjectivity of different aspects to take into consideration when designing refurbishment measures, such as values and money.

Traditional passive measures adopted in historical buildings use renewable energy sources, notably wind and solar, for heating, cooling and ventilation. Such measures include the design of patios and courtyards to improve building ventilation [25], the use of natural ventilation [26,27], double windows [28], and coloured reflecting mortars and tiles [29,30]. On the one hand, these measures can effectively maximize the intrinsic characteristics and behaviours of historical buildings, for example by using natural ventilation in heavyweight buildings for night pre-cooling in warm climates [26]. On the other hand, these approaches should consider aspects such as heritage value preservation and the energy embodied in the materials of the building, i.e., the energy required for their extraction, manufacturing, transportation, and during construction. For instance, Rosso et al. [29] demonstrated that the energy demand for cooling a building was decreased by using newly developed coloured reflecting mortars and tiles. However, the aesthetic impact of applying these new mortars should be evaluated against the potential loss of heritage values and the energy embodied in the original mortars and tiles. One could argue that if the historical tiles are not damaged, there is no reason to replace them, as their replacement will result in GHG emissions and a loss of heritage value. Effectively, the energy embodied in historical buildings is rarely considered in energy-retrofit strategies [31,32]. The adaptive reuse of heritage building materials can reduce GHG emission; nonetheless, historical buildings still need to be upgraded to be energy-efficient over their full life cycles [32]. Some assessments have compared the energy embodied in historical materials with materials that are more recent. However, such comparison should not be done on its own and needs to consider the historical and cultural value of the ancient materials. The adaptive reuse of historical buildings allow for preserving both the energy embodied in the material and the heritage values of the building [33].

Also, Litti et al. [34] investigated the replacement of historical windows with new ones as an additional measure to improve thermal insulation. The findings highlight that replacing windows does not necessarily allow for the largest energy savings over their full life-cycles, while their maintenance may result in comparable or more considerable savings [34]. In fact, the LCA methodology can be used to select solutions or materials using less energy and thus emitting less CO_2, however, the application of the LCA methodology is still in its infancy in the cultural heritage sector [35,36]. As a result of this gap in knowledge, Bertolin and Loli [37] developed a decision-support tool integrating a LCA approach within the framework of building conservation principles, nonetheless, they highlighted the need for further work in this direction given the complexity of this issue.

There are not only material aspects to consider when retrofitting historical buildings. Users' behaviour has an important role to play in mitigating climate change, as it determines the preference and choice for room temperature and ventilation, for instance [38], and can help target specific groups in carbon reduction strategies. Many studies highlighted the high energy saving potential derived from changes in user behaviour [39–43], which they estimated to range from 62 to 86% [40]. Human behaviour, however, can lead to a rebound effect in energy usage. Hens et al., [44] showed an example of such a rebound effect with the average indoor temperature of houses increasing after improving insulation. The increasingly energy-intensive way of life should also be considered when designing energy-efficiency policies and strategies by promoting lifestyles compatible with carbon reduction [45]. Occupants' behaviours in their use of energy can be approximated using variables such as type of dwelling or the Heating, Ventilation and Air Conditioning (HVAC) system they use, and thus help at targeting carbon reduction strategies to specific groups. Guerra Santin et al. [46] found that occupants living in non-detached dwellings or in houses where thermostats are installed consume less energy, for instance. This shows that more engagement with behavioural research is needed to identify opportunities for reducing GHG emissions [47].

Yarrow [48] investigated the perspectives of building professionals, planners and home owners in relation to not only the issue of climate change mitigation when refurbishing historical buildings, but also on the challenges involved in relation to the impact of energy efficiency improvements on the historical significance of heritage buildings. On the one hand, there is pressure to preserve the authenticity of the historic built environment, but, on the other hand, there is pressure to mitigate climate change. Considering the above options, it is clear that many mitigation choices involve physical alteration of heritage assets. This raises the question as to whether the mitigation solutions proposed in the literature are compatible with the heritage values and also with the traditional characteristics and behaviour of historical materials and structures. This is a complex issue and there are contrasting examples in the literature. For instance, Ascione et al. [49] developed a methodology to select measures to retrofit a historical building according to energy, environmental and economic indicators. Using that methodology, one of the solutions identified was the replacement of the historical windows with new double glazed ones, a solution that was proposed without considering the values of the heritage building. As an example of best practice, De Santoli et al. [50] focused on reducing the heat load of a historical building and, to deal with this issue, their proposed solution consisted of an air exchange system integrated with the existing architectural elements of the building, such as chimneys and fireplaces. By converting them to a new role, the solution remained compatible with the heritage values, and minimised changes to the building fabric. Webb [43] further stressed that energy retrofitting of historical buildings can be an opportunity to help preserve them for future generations. There is an inherent complexity in balancing the drivers and constraints of mitigation-related energy retrofitting of historical buildings [48]. There is a desire to improve internal comfort and to reduce operating costs, while the need to preserve heritage values can constrain mitigation actions; for example, Cornaro et al. [10] discounted the potential for interior wall insulation due to the presence of frescoes.

Overall, the literature on approaches to reduce GHG emissions in historical buildings focuses on presenting case studies on measures to improve the thermal performance of historical buildings, e.g., building envelope insulation and upgrading of heating and cooling systems. This is in addition to generic advice on reducing the environmental footprint of historical buildings through retrofitting, renewable energy generation on site, offsetting carbon emissions, managing waste and using water more efficiently, both from a technical point of view e.g., [8,51,52] and for informing buildings' owners and the public see [53–55]. This review highlights the paucity of studies on the challenges to overcome in the cultural built heritage sector to mitigate climate change. A broader picture is needed to inform and support decision making on the priorities to consider when promoting climate change mitigation in the cultural built heritage sector. A number of questions remain insufficiently addressed in the literature, notably, how do experts involved in the preservation of the cultural built heritage consider climate change mitigation? What are the enablers for implementing mitigation strategies, and what are the barriers to overcome? Answering those questions is essential for the development of mitigation measures and for the identification of future research directions. To date, most studies have used quantitative methods, except for Yarrow [48] who followed a qualitative approach. The current study uses a qualitative methodology involving interviews with experts, and the above literature review was conducted to provide background information for the interpretation of the interviewees' responses. To the author's knowledge, this is the first paper that identifies enabling and constraining factors as well as examples of best practice in mitigating climate change in the built heritage sector as a result of consultations with experts.

3. Methodology

The methodology used in this study consisted of qualitative semi-structured interviews with experts in the preservation and management of cultural heritage. In total, 45 interviews were conducted in the UK, Italy and Norway; three European countries with different climates and heritage typologies. The selection of case study sites in those three countries allowed for triangulation of information. These sites were also selected as they form part of a larger project led by the first author with research on vulnerability and adaptation to climate change risks reported elsewhere [56,57]. The number of interviews was based on the principle of saturation in qualitative research, i.e., when it became evident that there was redundancy in the interviewees' answers and no new theme emerged, no additional interviews were conducted. The interviewees were academics and researchers working in different universities and research centres, including experts involved in EU-funded projects focusing on climate change and cultural heritage (42%); practitioners working in organizations and institutions with a focus on the preservation of cultural heritage (27%), and managers, coordinators and professionals involved with UNESCO World Heritage Sites (WHS) (31%). Eighty percent of interviewees have more than 10 years of experience working on preserving cultural heritage and there were more males than women interviewees (Table 1).

Table 1. Characteristics of the interviewees: type of organization where they work, number of years working in the cultural heritage sector and gender (n = 45).

Type of Organization.	Number of Year Working in the Cultural Heritage Sector	Gender
Universities and research centres = 42%	1–9 years = 20%	Male = 58%
Governmental institutions = 27%	10–19 years = 30%	Female = 42%
Heritage sites = 31%	20–29 years= 30%	
	30–39 years= 18%	
	>40 years = 2%	

The interviewees were from diverse backgrounds and specializations, including anthropologists, archaeologists, architects, conservation scientists, geologists, biologists, managers and coordinators of heritage sites, sustainability officers and urban planners. The structure of the interviews was prepared in advance, but during the interviews the order and number of questions varied according to the interviewees' expertise and answers (i.e., the interviews were semi-structured). Introductory questions on the professional background of the interviewees were followed by a list of questions focusing on themes related to climate change mitigation in the cultural heritage sector: GHG emission reductions at heritage sites, improvement in the energy performance of historical buildings, sustainability of materials, and methods used during conservation practices. Ethical approval for this research was obtained through the University of the West of Scotland procedure. The interviews were audio recorded and then transcribed, analysed and coded using the NVivo software (Version 11, QSR International (UK) Limited, Daresbury, Cheshire, UK). The interviews were conducted in English in the UK and Norway and in Italian in Italy. The Italian interviews were not translated during the coding process but some quotes were translated for the purpose of displaying samples of interview quotations in Tables 2–4.

Table 2. Selected quotations in relation to the factors enabling climate change mitigation in the cultural built heritage sector.

Themes	Quotes
Economic factors	*"(Give) incentives (to) (...) people to make compatible (...) interventions."* (Academic)
	"Incentives, (...) funding programs (...) could help people to (...) (do) things that otherwise they cannot do." (Heritage site manager)
	"The state or municipality should provide incentives (and give funding) (...) (also) for non-listed buildings. (...) The gap between economic, cultural, historical and social value should be bridged." (Academic)
	"When things cost more to do in the right way we have to have the willingness to pay more." (Heritage site coordinator)
	"Recovering money (...) through energy saving." (Academic)
	"It is better to give work to people that live in the place and that can do maintenance, instead of bringing low quality windows from abroad." (Academic)
Legislation and regulations	*"(It) would be really helpful, particularly with tenements, if there was more legislation."* (Heritage site manager)
	"There has to be work on the development of guidelines in collaboration with the municipalities." (Academic)
	"You have to get it written in management plans. (...) We need (...) tools, action and activities to help them." (Academic)
	"(Regarding) making the buildings more energy efficient, (enforce) policy requirements when it is a public building." (Academic)
	"To report the carbon usage of the building (...) on how the building is energy efficient." (Academic)
	Normally buildings consume more energy compared (with) how they were designed. Post-evaluation needs to be performed and needs a regulatory structure." (Academic)
Sustainable refurbishment strategies	*"Focus on reuse and adaptability. (Avoid) building waste."* (Academic)
	"For the stonework (...) they sourced stone from a bridge that has been dismantled." (Member of governmental institution)
	"Use natural resources and materials." (Heritage site manager)
	"We rely on natural materials. We (...) promote the broader advantages in sustainability of the historic environment in terms of social-economic sustainability (...) local jobs, skills. (...) Instead of getting something (from abroad), we (engaged) a local firm. (...) Sustainable in terms of materials, (...) low carbon in terms of the local economy." (Member of governmental institution)
	"(Use) sustainable (products). Instead of mineral wool, there are wood-based products. (...) Traditional materials: wood, stone..." (Academic)
	"Every construction phase should consider the LCA (...) (this) is little applied to cultural heritage. Its conservation (...) should be done with low emissions." (Academic)
	"(Historical buildings) will not be passive houses nor be A+++. We need to understand what we mean by sustainability. If we change everything, losing the material and the energy used to produce it, we will have a better building in terms of thermal performance, but we lost much more grey energy. We need to evaluate the sustainability through 360 degrees." (Academic)
Sustainable transportation strategies	*"(Pedestrianisation) of historical city centres is a positive thing."* (Academic)
	"Sundays without cars helps in sensitising (the issue). (...) To go in the city centre with the bike ... " (Researcher)
	"Promote appropriate transport. (...) people do not actually think to cycle instead of taking a bus." (Sustainability officer)
Change in user behaviour	*"Encourage people to conserve energy, save resources, recycle and reuse. (...) Those things have to become the norm, rather that something special."* (Heritage site manager)
	"We achieved 30% of reduction in carbon emission over 4 years based on (...) fabric interventions (and) change in behaviour." (Member of governmental institution)
	"Wear another jumper instead of putting in double glazing." (Heritage site manager)
	"Run climate change and mitigation energy awareness. (...) Reduce the demand." (Member of governmental institution)
	"Work with owners and motivate them." (Academic)
	"Engage the citizenship." (Researcher)
Knowledge	*"It has very much to do with knowledge (...) (and) information"* (Academic)
	"Sharing the knowledge (...) contributing to research" (Heritage site manager)

Table 2. *Cont.*

Themes	Quotes
Energy compensation strategies	*"Face (the problem) at district or city level. (...) (Historical) buildings can be put in a grid of energy distribution at district level (and receive) the energy surplus"* (Academic)
	"The concept of trade-off. (...) New design with thermal inefficiency in some parts but with improved efficiency in others." (Academic)

Table 3. Barriers in mitigating climate change in the cultural built heritage sector.

Theme	Quote
Economic factors	*"There is very little incentive for certain building owners."* (Academic)
	"Sometimes we are asking people to do things in a more expensive way." (Architect)
	"Many times, (...) economic aspect prevail over social, historical and cultural value." (Academic)
Lack of regulation	*"Guidelines were created but (...) they will remain only suggestions if there is no political will to be stricter."* (Head of heritage site association)
	"Provide regulation to force (energy efficiency upgrade) of properties." (Academic)
	"The regulation plan says that existing buildings can be refurbished (...) but the character, materials and colour should be kept. (...) People should keep the material, (...) (make) repairs, (...) (but people) change them (...) interpreting the unclear text in regulation." (Academic)
	"Governments (...) are interested in devolution of regulation to local level (...) (including) cultural heritage protection or listed buildings consent (...). But, do the (local authorities) have the capacity, (interest, and) (...) knowledge? And, when (...) you ask who is looking at cultural heritage, (...) they would say, "oh, we could not afford it". (...) There is the need to have more written advice. (Member of governmental institution)
Value	*"The older the building is and the more value the heritage has the harder it is. (...) If the building is mid-20th century (...) (you can) get it in a high standard. But (...) (with) a castle there is a limit on how far you can go."* (Academic)
	"Insulating the outside of historical buildings in terms of authenticity and integrity (...) it is unthinkable. There is a balance with the cultural significance of the elements." (Academic)
	"I wouldn't like to (...) see solar panels all over beautiful old buildings. (They can) (...) be sited somewhere (else)." (Heritage site manager)
	"You never want to lose the original fabric, like crown glass or (...) timber sashes to put double glazing. (...) You never get character back again because you can't make crown glass anymore. (...) With traditional stone buildings (...) you don't want to alter the external fabric (...) (and) lose the (internal) original cornices and plaster work." (Heritage site manager)
	"They prefer to build new low emission buildings and demolish existing historical buildings. (...) If you look only at the environmental issues and not at the cultural (value) you are making wrong choices." (Academic)
	"Lack of information and education of the community on (...) heritage (value)" (Academic)
	"(When) refurbishing without taking into consideration the authenticity of the materials (...) (historic elements) are replaced (...) with totally different detailing." (Academic)
Material procurement and sustainability certification	*"Local materials should be more sustainable than (foreign), (...) (but) emission from shipping are not accounted in any assessment (and) it (results) more (sustainable) to buy from foreign countries."* (Sustainability officer)
	"Modern materials can travel all the world (...) (before to) come back to you." (Academic)
	"Little stock is produced. (...) (few) quarries (still) exist (...). You have to replace with stone (that) very often (...) comes from (abroad)." (Academic)
	"Resources from abroad (...) means that you are getting alien materials, there isn't a correct geological match." (Heritage site manager)
	"Local material (...) correspond better to (building) typologies and expression. (...) You (should) pay more for (...) (transportation) emissions." (Academic)
	"If you are replacing (elements) (...) are you calculating those carbon costs? (...) What about the destruction of what (...) has last seven hundred years? (...) It is a waste of money (...) (and) energy." (Heritage site manager)
	"This country calculates CO_2 only within the borders (...). So, (foreign) stone (...) has much smaller carbon footprint legally compared from (local) stone." (Academic)
	"(Sustainability assessments) are very good for new construction, (...) (but tricky with) refurbishment. (...) EPC (and) (....) building standards are really good for (new) or just refurbished buildings, but (there is) nothing on historic values. (Sustainability officer)
	"There are problems on how the (EPC) rating presumes the performance (...) we do not have data on the actual performance of the buildings." (Academic)

Table 3. *Cont.*

Theme	Quote
User behaviour	*"You make (a building) more energy efficient and theoretically you reduce the carbon emissions, then you find that they are using it in a way that it is not proper. Carbon emissions are made also from the users."* (Academic)
	"There is a cultural resistance to lighting from windows. (…) Electric artificial systems used when not necessary it is a big issue. People still think that lighting is low in consumption, but it is not." (Academic)
	"Everybody wears t-shirts (in offices). (…) We do not try to save the energy as hard as we could." (Heritage site manager)
	"The use-phase is important. (...) (heat pumps) were introduced to use less electricity, but now people have 25 degrees (indoor) (...). I do not know if they actually decreased the energy at the end. It is a rebound effect. You can have zones in your house (...), night and day regulation, (...) educate people (to) (...) put on a jumper in winter (…). (But) they just want to stay with their t-shirt. (…) People do not know why they should (do) it." (Academic)
Lack of knowledge	*"If we (use) (...) a modern material, it can cause increase of decay with formation of mould, for example. We need tests to know the consequences of their application."* (Academic)
	"We need to understand which are the characteristics and the behaviours (of ancient buildings) and work with them." (Academic)
	"We need skills of preservation of (old) materials. (...) (and) use the same methods (…) used when they were built, not look only at today's methods (…) and guidelines for energy efficiency." (Heritage site coordinator)
	"The sash and case windows were designed for having air conditioning. (...) A lot of historical buildings have this inherent ability. (...) There is the need to understand the capacity of these buildings to cope with the external environment and to maintain (comfort)." (Member of governmental institution)
	"If people's opinion is that new is always better, they are going to replace (elements). (...) If you promote to keep them (...) that would help. But the (market is) (...) advertising for new windows. Heritage managers are not considered experts in energy efficiency." (Academic)
	"People do what they want (...) entrepreneurs put new insulation (...) (which cause) problems with air tightness compromising your building." (Academic)
Loss of traditional skills	*"Traditional skills are going to be lost. If I have a technician that just mount PVC windows instead of a craftsman (...) able to build a wooden window or to fix an ancient door, we are losing something."* (Academic)
	"Other important resources are the technicians. We do not have enough people who know traditional skills" (Academic)
	"Some craftsman know about (historical buildings behaviour) but a lot (...) use the same techniques for new buildings in the old ones; (…) the building industry does not take this into consideration." (Academic)
Incompatible solutions	*"Changing the behaviour of the building will cause inside a series of problems."* (Academic)
	"We need to be careful in adding insulation (...) (in) buildings. You can upset their balance, (...) blocking natural ventilation and causing cold bridges." (Heritage site manager)
	"You want to (...) increase energy efficiency, but people are not worried about humidity and condensation causing deterioration in the fabric." (Member of governmental institution)
	"(In) wooden buildings (...) there would be vapour condensation somewhere, it is very important to (...) ventilate (....). Damp barriers (...) prevent the house to breathe (....) (and) cause decay. People do not understand (or know about building physics)." (Academic)
	"Some levels of comfort are not sustainable for some heritage buildings. (...) Heating and cooling inside churches can increase decay." (Academic)
	"The energy strategy (in refurbishments can be) contradictory. (...) (A refurbished building) had an overheating problem (...) because of the (...) IT equipment. (...) They (internally) insulated a wall with a huge thermal mass, (that) now cannot absorb the heat from computers, and they have to activate air conditioning." (Member of governmental institution)
Diversity	*"You have to take each building or set of buildings. When were they built? How? (...)For what? What use can be made of it now?"* (Academic)
	"That would be a question to raise in each building and the answer would be different from building to building. It is hard to generalize." (Academic)

Table 4. Best-practice examples of cultural heritage refurbishment measures to mitigate climate change, as proposed by the interviewees.

Theme	Quote
Insulation	*"Internal insulation, heat pump, soil insulation and roof insulation, improving the windows..."*
	"Maintain the wooden windows. (...) Putting insulation in the roofs, putting double glazing in the windows, using heat pumps..."
	"Insulation of floors, ceilings and roofs..."
	"Put secondary glazing on historic windows (...) removable secondary glazing with magnets (...) double glazing in winter and in summer you can take it off."
	"Put insulation under the floor, in the ground floor, towards the basement, in the roof. (...) Make the building more airtight. Put secondary glazing or double glazing in historic windows without replacing them. (...) Make zones. Halls in the entrance. (...) Insulation in certain walls (...) that are re-built.
Insulation and ventilation	*"Installing secondary glazing in windows to reduce the heat loss, improving the insulation in walls, floors and ceilings. Reducing drafts. (...) Maintain the ventilation in historical buildings (...) in roofs. (...) Closing curtains and having shutters."*
	"Have shutters, heavy curtains, and secondary glazing. (...) Make sure that there is good ventilation. Add insulation where appropriate."
	"Insulate in the inside (only if adequate) (...) provide the right ventilation (...) make sure that you have got no condensation (...) new curtains or tapestry on the walls. (...) Using the blinds, using curtains (...). Insulate under floor."
Ventilation	*"Mixed mode or hybrid ventilation system, in some cases natural ventilation supported by mechanical ventilation. (...) Mechanical devices to improve the natural ventilation of buildings in a controlled way."*
	"Passive ventilation. (Avoid) air conditioning. (...) A lot of buildings (...) had passive ventilation, (...) like chimneys and louvers (...) (that) provide thermal comfort with less energy."
Lighting	*"Good light management process."*
Lighting and insulation	*"Changing light bulbs, repairing windows, secondary glazing, new lamping, loft insulation, cavity wall insulation."*
Heating	*"Do not heat the building but heat the people."*
Heating, lighting, monitoring.	*"Install LED bulbs. (...) Change the windows putting up films for the summer shading, upgrade the roof. Improve the cooling and lighting systems. (...) Monitoring the environmental conditions all year."*

The interviews transcripts were coded using the categories: "enablers", "barriers" and "best practice". Subsequent to coding, the transcripts were used to search for patterns in the data. This search for patterns in the coded interview transcript led to the identification of themes representing the factors enabling and those impeding climate change mitigation in the built heritage sector was then determined by analysing the coded interview transcripts and are displayed in Figure 1.

Examples of quotes from the interviewees in relation to each of the identified themes are shown in Tables 2–4 to explain the selection of specific themes and the categorisation of the information provided by the interviewees. These representative quotes were carefully selected and shortened to reduce the word count without compromising their meaning. The use of direct quotations allows the reader to have a better understanding of the respondents' perceptions about climate change mitigation. The results do not present commonalities or differences amongst the interviewees, as the purpose of this paper is to highlight the factors enabling or constraining climate change mitigation in the cultural heritage sector, including examples of best practice for its implementation, as well as to inform decision-makers and to identify research needs. Figure 2 summarises the steps involved in this qualitative methodology:

Primary coding	Secondary coding
ENABLERS	Economic factors
	Legislation and regulations
	Sustainable refurbishment strategies
	Sustainable transportation strategies
	Change in user behaviour
	Knowledge
	Energy compensation strategies
BARRIERS	Economic factors
	Lack of regulations
	Value
	Material procurement and sustainability certifications
	User behaviour
	Lack of knowledge
	Loss of traditional skills
	Incompatible and contradictory solutions
	Diversity
BEST PRACTICE	Insulation
	Ventilation
	Lighting
	Heating
	Monitoring

Figure 1. List of primary and secondary coding.

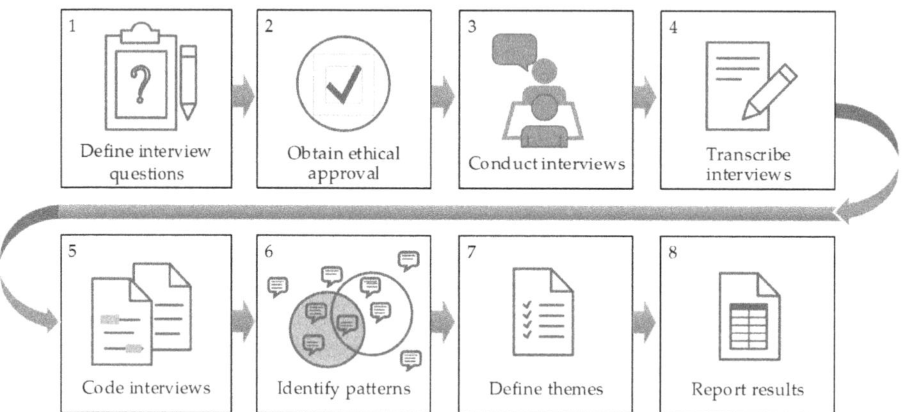

Figure 2. The qualitative methodology used in the current research.

4. Results and Discussion

The investigation first examined the interviewees' perceptions to identify the factors that enable and those that act as barriers to climate change mitigation in the cultural built heritage sector. During the interviews, the participants addressed issues such as reducing carbon emissions and the transition to a low carbon economy, improvements in the energy performance of historical buildings, the preservation of heritage values, the sustainability of materials and methods used for refurbishing and restoring heritage assets.

4.1. Enablers of Climate Change Mitigation in the Cultural Built Heritage Sector

The interviewees were positive about adapting the cultural built heritage to mitigate climate change and identified a number of factors that enable this. These factors were grouped into six themes: 'economic factors', 'legislation and regulations', 'sustainable refurbishment strategies', 'sustainable transport strategies', 'user behaviour', 'knowledge' and 'energy compensation strategies' (see Table 2). Figure 3 shows the number of interviewees mentioning each of the identified enablers. For example, 13 interviewees, i.e., the 29% of the total number of interviewees, mentioned 'economic factors' as an enabling factor to climate change mitigation in the cultural heritage sector.

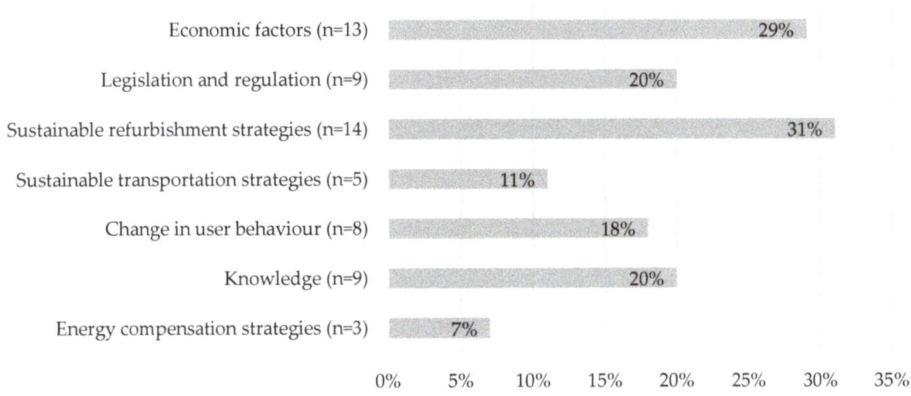

Figure 3. Percentage (and number) of interviewees mentioning each of the different enablers of climate change mitigation. (*n* is the number of interviewees mentioning the enabler).

4.1.1. Economic Factors

The interviewees perceived as important the economic factors that enable the implementation of strategies to mitigate climate change. Government incentives was the most mentioned measure by the interviewees. Such incentives are used to encourage people to make interventions compatible with the heritage assets instead of choosing market-forced solutions that are often not compatible with the heritage values. The interviewees found that the wider (non-expert) public can be reluctant to restore traditional windows due to the high costs involved and the belief that the restored windows will be less energy efficient than alternative new double-glazed windows. However, as mentioned above, the results of previous research concluded that window replacement is not necessarily a better energy saving measure. Considering the energy used in the life cycle of the windows, restoring, instead of replacing the windows, can be a better choice [34]; in addition, the energy cost related to the transport of the new elements is avoided. The interviewees also emphasised that restoration efforts compatible with the heritage building should be encouraged with the help of incentives so that people are willing to pay more for them. This is particularly the case when the money invested in the refurbishment efforts can be recovered in the long-term from the energy saved subsequent to the refurbishment efforts. The interviewees suggested restoring and maintaining the original historical materials as an opportunity for improving local economic sustainability against the importation of foreign materials and for maintaining as well as increasing traditional expertise locally.

4.1.2. Legislation and Regulations

According to the interviewees, having more legislation, regulations and guidelines would facilitate climate change mitigation in the cultural heritage sector. New regulations were requested not only at the international, national and regional level, but also in municipalities and at cultural heritage sites. In fact, it was even suggested to include climate change mitigation in site management plans, but also to increase policy requirements for mitigation actions in public buildings. A specific request by the interviewees was the need to improve the regulatory structure behind carbon usage reports and post-performance evaluations in order to better monitor the energy consumed in refurbished buildings during the use-phase and to compare it with the energy usage that was predicted prior the refurbishment. Consistent with this information gathered from the interviewees, there is agreement in the literature that ambitious policies can significantly reduce the energy use in buildings [58], and that for successful mitigation in the cultural heritage sector, policy actions at both the regional and local levels are needed [7,59]. In addition, Yarrow [48] highlighted a potential conflict between legislation promoting preservation of cultural heritage and that requiring a reduction of energy consumption in buildings, hence the difficulty of applying general policies to historical buildings.

4.1.3. Sustainable Refurbishment Strategies

The interviewees suggested a number of actions aimed at improving sustainability and thus at reducing the GHGs emitted by the cultural built heritage sector. Examples of sustainable actions in relation to the refurbishment of historical buildings include the selection of environmentally friendly and natural materials, especially if originating from local sources, recycling and reusing materials, avoiding the use of harmful chemical substances and minimising the generation of building waste. However, sometimes it is difficult to understand whether one way to refurbish a historical building is more sustainable than another is. For this reason, the sustainability of the materials and refurbishment actions should be assessed using the LCA methodology prior to the implementation of the refurbishment efforts. This would provide a better understanding of the energy and CO_2 emissions associated with a specific intervention. Such an approach, however, should also consider the embodied energy in the historical materials that are proposed to be replaced. The latter is consistent with Webb [43] who mentioned the lack of consideration of the embodied energy of historical buildings during retrofitting. Hence, further work is required on the evaluation of the sustainability of the materials and methods used during refurbishment in the built heritage sector, as the LCA approach has been applied mostly on new construction rather than on historical buildings [48], a gap in knowledge which Bertolin and Loli [37] also had previously highlighted. In addition, when applying the LCA approach to the heritage sector, the historical and cultural values need to be considered and added as a parameter in the evaluation of the solutions to be adopted. These values are often neglected and priority is often given to new technologies and to materials targeted at energy reduction [37], without considering the sustainability of the cultural built heritage sector in a holistic way as mentioned by an interview in the following quote: "We need to evaluate (...) sustainability through 360 degrees." (Table 1).

4.1.4. Sustainable Transportation Strategies

The interviewees also emphasised the need to improve environmental sustainability at cultural heritage sites by reducing CO_2 emissions through the promotion of sustainable transport to and from the sites as well as within the sites. This can be accomplished, for example, by promoting cycling and public transportation, as well as by closing historic town and city centres to domestic vehicles. The latter can also contribute to decreasing air pollution and thus improving air quality, as well as reducing the associated decay on monuments and the facades of historical buildings [60].

4.1.5. Change in User Behaviour

User behaviour has a significant impact on the effectiveness of mitigation measures adopted on historical buildings. The interviewees recommended to encourage users, e.g., occupants and visitors, to behave in a more sustainable way and mentioned the need to raise awareness through engagement to ensure that mitigation is effective. Users should be educated in energy and water saving measures, reusing and recycling materials, and decreasing their waste and emissions. This can start through simple things such as wearing a jumper inside the building in the winter and thus reducing the use of heating, instead of wearing only a t-shirt, for instance. Berg et al. [38] emphasised the importance and potential of user-driven energy efficiency measures in historical buildings and that user behaviour should always be taken into account, as the occupants play a central role in the day-to-day management of the building. The Intergovernmental Panel on Climate Change (IPCC) [58] further highlighted that in developed countries, behavioural change could decrease energy use by up to 20% in the short-term and by up to 50% by the middle of the 21st century.

4.1.6. Knowledge

Some interviewees suggested more knowledge through research and dissemination as a factor enabling mitigation in the cultural heritage sector. This is consistent with the literature, which emphasises the role of new knowledge and skills to develop new sustainable strategies for heritage conservation to face climate change uncertainty [51].

4.1.7. Energy Compensation Strategies

Some interviewees suggested the application of energy compensation schemes. This is where parts of a system that are inefficient in their use of energy can benefit from the energy surplus produced by the efficient parts. In relation to cultural heritage, this can be done at different scales: city district and building. In the first case, different buildings in the same district can share energy. For instance, the energy needed for historical buildings can be supplied by the surplus energy produced by new energy efficient buildings or from off-site renewable sources of energy. GHG emission reductions through such a trading allowance is a cost-saving measure that was proposed by Cassar [51]. In the second case, one interviewee mentioned the possibility to meet the energy requirements for parts of a historical building where no refurbishment is allowed with the energy saved by improving the thermal performance of other parts of the buildings where modifications are permitted.

4.2. Barriers to Climate Change Mitigation In The Cultural Built Heritage Sector

In addition to the factors found to enable climate change mitigation in the cultural built heritage sector, the interviewees also identified a number of barriers. These barriers were grouped into nine themes: 'economic factors', 'lack of regulation', 'value', 'material procurement and sustainability certification', 'user behaviour', 'loss of traditional skills', 'lack of knowledge', 'incompatible solutions' and 'diversity'. Figure 4 shows the number of interviewees mentioning each of the identified enablers as well as their percentages. Eleven interviewees, i.e., the 24% of the total number of interviewees, mentioned 'economic factors' as a barrier to climate change mitigation in cultural heritage sector. Table 3 provides examples of quotes from the interviewees in relation to each enabling factor.

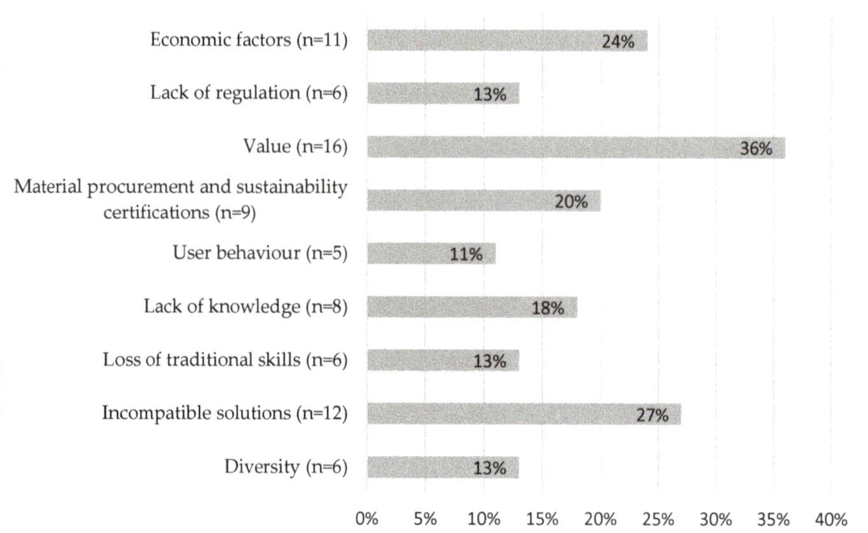

Figure 4. Percentage of interviewees mentioning each barrier (*n* is the number of interviewees mentioning the barrier).

4.2.1. Economic Factors

The interviewees perceived economic factors as an important barrier to climate change mitigation in the cultural built heritage sector as restoring historical buildings is expensive. Thus, financial resources can help with conservation efforts when they are limited or not available. The lack of money to restore cultural heritage can also have negative consequences on the inherent historical, cultural and social values embedded in heritage sites. For example, an interviewee stated: "There is an economic aspect that prevails over the social, historical and cultural value." (Table 2). The adoption of cheaper solutions and in particular the loss of original material can influence the historical and cultural values of the site, as well as weaken the identity of a place. The economics of conservation is therefore an important criterion to consider when retrofitting historical buildings [43].

4.2.2. Lack of Regulation

A significant barrier perceived by the interviewees is the lack of regulation when adapting cultural heritage to mitigate climate change. The interviewees highlighted that stronger regulations are required to force the owners of heritage assets to adapt properly. An interviewee mentioned the example of a UNESCO WHS, where detailed guidelines were carefully designed and provided to the local community, but people did not follow them due to lack of enforcement. An interviewee from another WHS mentioned misinterpretation of the regulations as an issue, where the components of historical buildings were replaced with similar substitutes, albeit made of different materials characteristics, varying in shape and colour from the original components. Such substitution can decrease the authenticity of a heritage site. There are parallel examples between the information provided by the interviewees and case study examples reported in the literature. Yarrow [48], for instance, reported the concern of a conservation officer regarding a retrofitting where the building owners substituted the historical components for new ones claiming that they 'looked the same'. Devolution of the regulations to the local level was also mentioned as a problem by one interviewee, because, in some cases, a lack of financial and human resources put cultural heritage preservation in second place. The lack of protocols and guidelines to help heritage managers adapt cultural heritage assets to climate change was also highlighted by Sabbioni et al. [8].

4.2.3. Value

Some interviewees perceived preserving the values of cultural heritage as a barrier to climate change mitigation, because it limits the refurbishment options to make historical buildings more energy efficient. This is because the measures adopted when retrofitting heritage assets can affect the heritage value, notably when historical elements are removed and replaced. Preserving the values of cultural heritage on the one hand and the implementation of measures to reduce the carbon footprint of heritage assets on the other hand should always be carefully balanced in any refurbishment project and, in this respect, buildings of diverse ages and listing category should be considered differently. The interviewees highlighted that preserving heritage value, authenticity and integrity should be prioritized when considering adaptation measures and strategies, as once the historical materials are removed and destroyed they will be lost forever. For this reason, the interviewees suggested alternatives to refurbishing buildings of high heritage value, such as integrating them as part of district energy networks, as described in Section 4.1.6, with the curbed CO_2 emissions from retrofitted buildings used to offset the higher energy used in historical buildings.

One interviewee further added that it would be beneficial to have examples of public building refurbishment that preserve the value, integrity and authenticity of the heritage as models for others to follow: *"Long-term adaptation planning has to be done first in public buildings to be an example for the private."* (Table 3). Another interviewee representing a government organisation recommended selecting adaptation measures that would not conflict with the heritage value: *"Where we had (...) a potential for losing the values, we just do not go there. We would rather do something else."* An example of the latter is restoring historical wooden windows by inserting a secondary glazing with magnets, which can be removed during the warmer season, rather than replacing the windows with new ones. When refurbishing historical buildings, evidence is needed to justify retrofitting changes that influence the value, integrity and authenticity of the building. If restoration impacts on the value of those building, their historical meaning could be lost [51]. Historic Environment Scotland (HES), a governmental institution in charge of preserving cultural heritage in Scotland advises on the unrealistic expectation that historical buildings can reach energy efficiency levels similar to new constructions but also advises for adaptive solutions that preserve the value, identity and significance of those assets to the community [61]. Furthermore, it should be emphasized that if the refurbishment of historical buildings can be seen as a threat to the heritage values, retrofitting can also constitute an opportunity to protect buildings, for example to heritage buildings in disuse or to those with minimal maintenance carried out [43].

4.2.4. Material Procurement and Sustainability Certification

The interviewees identified material procurement used during refurbishment of cultural heritage buildings as a barrier to climate change mitigation, as well as weaknesses in building sustainability certifications. With globalization, building materials are increasingly imported from other countries and, during the production process, they often travel from various locations around the world before reaching the construction site. The more a material travels, the more GHG emissions are derived from its transportation. However, the interviewees highlighted that in some cases, energy calculations and sustainability certifications do not consider the energy associated with transportation out of the country where the building is located. This can put foreign building materials at an advantage with regard to costs and certification, and this can influence the lack of locally sourced material during the refurbishment of historical buildings. The interviewees emphasized that the use of foreign materials not only increases GHG emissions, but may also not have the same characteristics as the original materials. There is also a lack of consideration of the energy embodied in the historical materials in energy evaluations. In fact, it was stated that historical buildings are not considered in performance qualification and building standards and that energy evaluation software are not accurate in calculating the energy used by historical buildings.

The interviewees also reported a lack of consideration of the physical behaviour of heritage buildings and of the vapour permeability of historical material and an inaccuracy of Energy Performance Certificates (EPC) that make assumptions about energy use but do not reflect the actual performance of historical buildings. Incomplete Environmental Product Declarations (EPD) can also contribute to this issue [62], and, consistent with this, Cassar [51] stressed that some sustainability certification tools do not use a holistic approach; they are based on relative energy targets and they exclude social and economic sustainability aspects. In addition, Webb [43] emphasised the lack of consideration of embodied energy during the retrofitting of historical buildings, stressing the need for improved simulation tools and software to reduce the gap between simulations data and real building performance.

4.2.5. User Behaviour

User behaviour is considered a barrier to climate change mitigation because certain attitudes can minimise or even nullify the refurbishment efforts aimed at decreasing the use of energy by the building. The interviewees identified resistance to change and to the adoption of sustainable behaviours, for example, leaving the lights on during the day when there is high luminosity from the window, wearing light clothes in the winter and turning the thermostat up to a level beyond comfort level, to name a few. An interviewee also mentioned the rebound effect, which is a decrease in the expected energy savings from a specific building alteration because of behavioural change leading to an increase in energy consumption. For instance, in buildings were measures were introduced to use less electricity by installing heat pumps, people further increased their energy consumption by setting room temperature to a higher level on their thermostat. Sabbioni et al. [8] even proposed a change on a regulatory level to induce a modification in user behaviour.

4.2.6. Lack of Knowledge

The interviewees identified a lack in knowledge as a barrier to mitigate climate change in the built heritage sector. More research testing the use of modern materials during the refurbishment of historical buildings and on the traditional behaviours and characteristics of historical building is needed. Some interviewees also identified the need for more knowledge transfer to users on the topic of climate change mitigation in the field of cultural heritage in a way to make them more aware and capable of making choices compatible with the traditional behaviours and values of historical buildings.

4.2.7. Loss of Traditional Skills

Another significant barrier identified by the interviewees is the progressive loss of traditional skills and techniques. Given the common incompatibility of modern solutions with historical buildings and materials, this loss can threaten heritage preservation. The need to pass on knowledge and skills to the next generation of heritage professionals, and the importance of teaching traditional skills and techniques, and to understand how they relate to modern construction materials is emphasised in the literature, as well as methods on how to develop suitable conservation techniques appropriate with the nature of older buildings [51,63].

4.2.8. Incompatible Solutions

When refurbishing historical buildings, the adoption of solutions that are not compatible with the behaviour of historical materials can accelerate their degradation. For instance, the interstitial condensation between layers made of different materials. The traditional behaviour of the historical buildings, their thermal-balance, the natural ventilation of their indoor environments, and the ability of some historical materials to allow vapour movement are characteristics that must be taken in consideration during a refurbishment project. More research is needed on the evaluation of market-based solutions, which are usually designed for new structures but that are also applied to historical buildings. Historical buildings have more sophisticated bioclimatic properties than modern constructions, and new materials that are incompatible with the physical behaviour of the historical ones can lead to an increase in decay. Retrofitting measures can alter the moisture balance of historical buildings, affecting, for example, their breathability, leading to an increase in the deterioration of historic materials [43,48]. Those buildings are designed to have controlled thermal comfort based on the natural unconditioned climate, environmental and site conditions rather than using machine-driven systems. The development of adaptation strategies should therefore consider the balance between decay reduction and energy consumption improvement of historical buildings [64,65].

4.2.9. Diversity

Some interviewees mentioned that historical buildings are too diversified for adaptation solutions to be generalised and that adaptation has to be considered on a case-by-case basis. The question remains as to whether it is possible to generalize adaptation options to decrease the energy consumption of historical buildings or if each individual cultural heritage asset is too specific. More research is needed on specific classifications, or typologies, of heritage buildings to understand which possible adaptation solution can be proposed by comparing adaptation measures used in similar situations.

4.3. Best Practice Examples

The last part of the investigation focused on the identification of best-practice examples in mitigating climate change, as expressed by the interviewees. Examples of best practices include refurbishment measures to increase the energy efficiency of historical buildings in relation to upgrading the insulation, ventilation, lighting and heating, as well as the monitoring of climatic conditions (Table 4). The majority of these measures are consistent with the solutions proposed in the literature and raised in Section 2 of this paper.

4.3.1. Insulation

The interviewees suggested insulating floors, ceilings, roofs and walls, and to reduce the loss of heat through windows by adding secondary glazing and using curtains, shutters and blinds. Examples of insulation systems used in literature are shown in Section 2 of the current paper e.g., [14–16].

4.3.2. Ventilation

The interviewees emphasised the need for ventilation in historical buildings to avoid condensation, which can be done by using mixed mode or hybrid ventilation systems, with mechanical devices improving natural ventilation of buildings in a controlled way, or by passive ventilation using traditional systems such as wind chimneys and louvers. Consistent with this, the literature stresses the importance of systems to control the relative humidity and the ventilation of the indoor environment of cultural heritage buildings to decrease energy consumption while also reducing the risk of degradation [17], and on the potential of chimneys for providing ventilation in traditional buildings [53].

4.3.3. Lighting, Heating and Monitoring

The interviewees also suggested improving lighting and heating systems while monitoring climatic conditions throughout the year. The literature presented in Section 2 backs these recommendations.

5. Conclusions

Mitigating climate change by reducing GHG emissions is now urgent to prevent future—and reduce current—danger to cultural heritage resources. This needs to be accomplished, in part, by reducing the energy used by cultural built heritage assets, notably through improvements in their energy efficiency. Organizations such as UNESCO, ICOMOS, ICCROM, IPCC and the European Commission promote actions to cut GHG emissions in the cultural built heritage sector. This paper analysed the perceptions of cultural heritage experts on the issue of adapting the cultural built heritage to mitigate climate change. Specifically, it reports on the factors enabling and those acting as barriers to mitigate climate change, as perceived by the managers of heritage sites and experts working on the preservation of cultural heritage in universities, research centres and governmental institutions.

A common view amongst the interviewees was that climate change mitigation in the heritage sector is necessary but challenging. Most research accomplished to date investigated measures to mitigate climate change by reducing the energy consumption of cultural heritage buildings, but there is limited research on the identification of the challenges to overcome in this regard. This paper provides a better understanding of what needs to be provided and prioritized for the mitigation of climate change in the cultural built heritage sector to take place. In summary, this study identified the following barriers constraining climate change mitigation in the field of cultural heritage: economic factors, lack of regulation, heritage values as it can limit the type and scope of refurbishment, inadequacies in material procurement and sustainability certifications, inefficient use of energy due to building occupants' behaviour, lack of knowledge, loss of traditional skills, the adoption of solutions incompatible with the assets and hence causing further damage, diversity of heritage resources and hence the difficulty to identify solutions that are fit for all. The factors enabling climate change mitigation that could help to overcome some of those identified barriers include economic resources and incentives, legislation and regulations, sustainable refurbishment strategies, sustainable transportation strategies, change in user behaviour, knowledge, and energy compensation strategies. Figure 5 provides a graphical summary of the factors enabling and constraining climate change mitigation in the field of cultural heritage.

This research emphasised that mitigating climate change in the cultural built heritage sector is a complex issue and requires a holistic approach for the identification of sustainable strategies to reduce GHG emissions. In this regard, ambiguities still remain on the consideration of heritage values and the energy embodied in historical materials, as well as the potential of traditional passive measures when adapting heritage buildings to mitigate climate change. This study improves our qualitative understanding of the key themes raised by stakeholders and can be used as a basis for further research seeking to quantify the effectiveness of best practice in climate change mitigation in the cultural heritage sector. Specifically, further research should be developed on adaptation strategies that consider the balance between reduction of decay and improving the energy consumption of historical buildings [64]. More research should also be conducted on the use of traditional solutions for improving energy efficiency, and on the compatibility (or lack) of new technologies and materials with the historical ones. Historical buildings work differently from modern buildings. Further research is required on the re-use of traditional passive measures such as wind chimneys, greenhouses, passive ventilation and heating, to name a few, and thus avoiding carbon intensive energy consumption, as well as on the promotion of sustainable natural resources, e.g., wind and sunlight. This could be done by understanding the level of energy efficiency achievable by different types of heritage (e.g. ancient castles and monuments versus non-ancient historical assets) in relation to their values, integrity and authenticity.

Monitoring before and after retrofitting and/or refurbishment works should be encouraged to estimate properly the energy consumption and the energy payback period from the use of renewable sources post intervention. This includes the need to rectify the current energy certification process, which does not consider the energy used to transport materials when historical materials are replaced by new ones in their calculations; the need for LCA evaluations before refurbishing historical buildings to understand the impact of the proposed solutions on the environment; and the need to promote the use of natural materials to avoid toxic chemicals.

A decrease in energy consumption also needs to be promoted through changes in user behaviour (e.g. wearing appropriate clothes and heating less, reducing the heating of rooms when not used, use more natural illumination and less electricity), which can be implemented with the use of information and communication technology (ICT). More knowledge needs to be disseminated and collaboration between different countries and at different levels should be encouraged, from the governmental to the end user. Including increasing awareness and cooperation between heritage organizations, governmental institutions, research centres and academia.

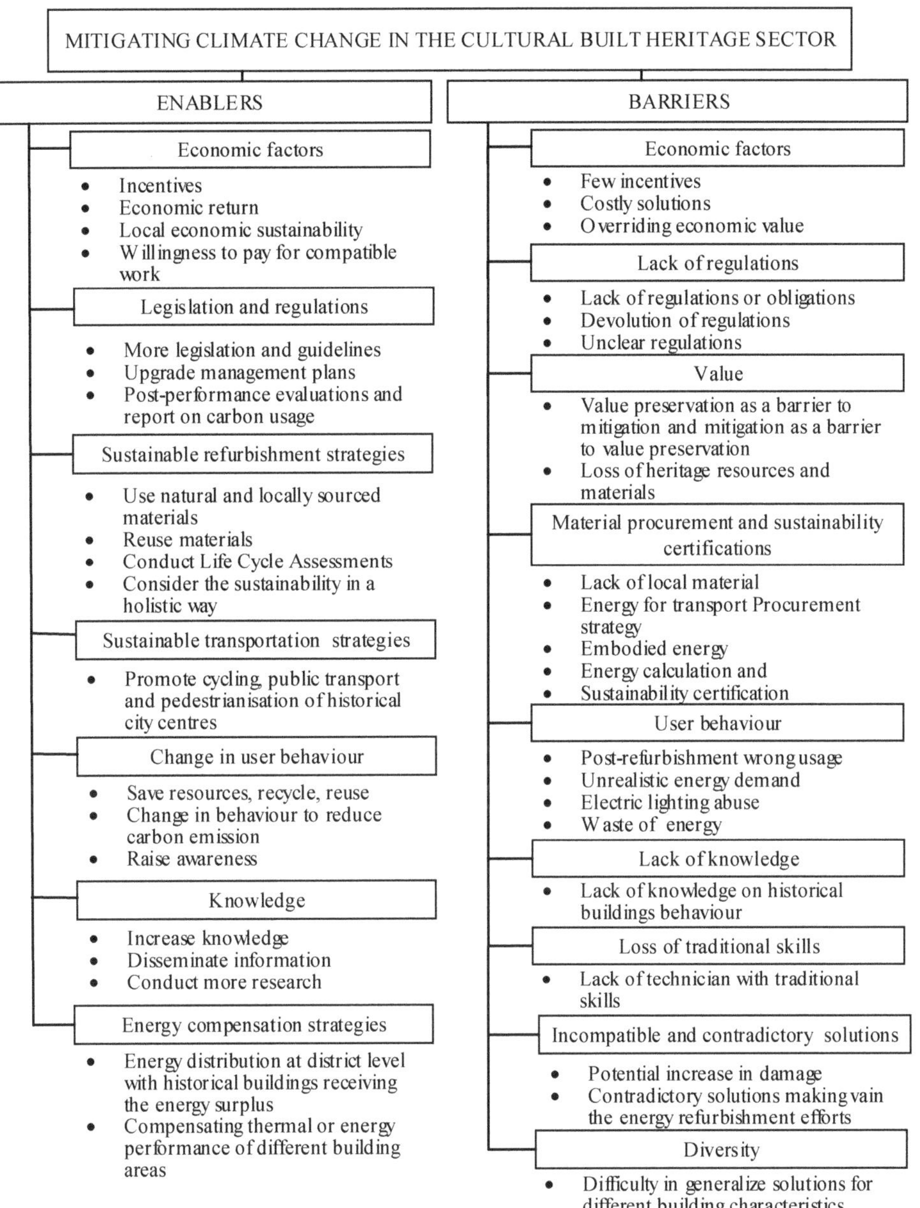

Figure 5. Summary of the research findings.

Author Contributions: Conceptualization, E.S., A.G., C.B. and J.H.; Data curation, E.S.; Formal analysis, E.S.; Funding acquisition, E.S., A.G. and J.H.; Investigation, E.S.; Methodology, E.S. and A.G.; Project administration, A.G.; Resources, J.H.; Supervision, A.G., C.B. and J.H.; Validation, C.B.; Visualization, E.S.; Writing—original draft, E.S.; Writing—review & editing, A.G., C.B. and J.H.

Funding: The University of the West of Scotland provided a university studentship to support the doctoral studies of the first author. The Postdoctoral and Early Career Researcher Exchanges (PECRE) funding scheme of the Scottish Funding Council and awarded by the Scottish Alliance for Geoscience, Environment and Society (SAGES) supported the academic exchanges of the first author to Italy and Norway.

Acknowledgments: We thank all the interviewees for their time and information, and Historic Environment Scotland for the grey literature provided. We are also grateful to The Institute of Atmospheric Sciences and Climate, National Research Council of Italy (ISAC-CNR) in Bologna and the Norwegian University of Science and Technology (NTNU) in Trondheim where two academic exchanges were conducted, and to the Scottish Funding Council via the Scottish Alliance for Geoscience, Environment and Society (SAGES) who funded those two academic visits.

Conflicts of Interest: The authors declare no conflict of interest.

References

1. European Commission. *Communication from the Commission to the European Parliament and the Council. The Road from Paris: Assessing the Implications of the Paris Agreement and Accompanying the Proposal for a Council Decision on the Signing, on Behalf of the European Union, of the Paris Agreement Adopted Under the United Nations Framework Convention on Climate Change*; European Commission: Brussels, Belgium, 2016.

2. EU. Energy performance of buildings. Available online: https://ec.europa.eu/energy/en/topics/energy-efficiency/buildings (accessed on 21 August 2018).

3. European Commission. *Communication from the Commission to the European Parliament, the Council, the European Economic and Social Committee, the Committee of the Regions and the European Investment Bank*; European Commission: Brussels, Belgium, 2015.

4. European Union. *Directive 2010/31/eu of the European Parliament and of the Council of 19 May 2010 on the Energy Performance of Buildings*; European Union: Brussels, Belgium, 2010.

5. UNESCO. *World Heritage in Europe Today*; United Nations Educational, Scientific and Cultural Organization: Paris, France, 2016.

6. Colette, A. *Climate Change and World Heritage. Report on Predicting and Managing the Impacts of Climate Change on World Heritage and Strategy to Assist States Parties to Implement Appropriate Management Responses*; World heritage report; UNESCO World Heritage Centre: Paris, France, 2007; Volume 22.

7. Cassar, M. Principles of mitigation and adaptation of cultural heritage to climate change. In *Climate Change and Cultural Heritage, Proceedings of the Ravello International Workshop, 14–16 May 2009 and Strasbourg European Master-Doctorate Course, Strasbourg, France, 7–11 September 2009*; Lefèvre, R.-A., Sabbioni, C., Eds.; EDIPUGLIA: Bari, Italy, 2010; pp. 43–47.

8. Sabbioni, C.; Brimblecombe, P.; Cassar, M. *The Atlas of Climate Change Impact on European Cultural Heritage. Scientific Analysis and Management Strategies*; Anthem Press: London, UK, 2010.

9. Hambrecht, G.; Rockman, M. International approaches to climate change and cultural heritage. *Am. Antiq.* **2017**, *82*, 627–641. [CrossRef]

10. Cornaro, C.; Puggioni, V.A.; Strollo, R.M. Dynamic simulation and on-site measurements for energy retrofit of complex historic buildings: Villa mondragone case study. *J. Build. Eng.* **2016**, *6*, 17–28. [CrossRef]

11. Dalla Mora, T.; Cappelletti, F.; Peron, F.; Romagnoni, P.; Bauman, F. Retrofit of an historical building toward nzeb. *Energy Procedia* **2015**, *78*, 1359–1364. [CrossRef]

12. Franco, G.; Magrini, A.; Cartesegna, M.; Guerrini, M. Towards a systematic approach for energy refurbishment of historical buildings. The case study of albergo dei poveri in genoa, Italy. *Energy Build.* **2015**, *95*, 153–159. [CrossRef]

13. Righi, A.; Mora, T.D.; Peron, F.; Romagnoni, P. Historical buildings retrofit: The city hall of the city of motta di livenza (tv). *Energy Procedia* **2017**, *133*, 392–400. [CrossRef]

14. Berardi, U. The benefits of using aerogel-enhanced systems in building retrofits. *Energy Procedia* **2017**, *134*, 626–635. [CrossRef]

15. Zhou, X.; Carmeliet, J.; Derome, D. Influence of envelope properties on interior insulation solutions for masonry walls. *Build. Environ.* **2018**, *135*, 246–256. [CrossRef]

16. Nardi, I.; de Rubeis, T.; Taddei, M.; Ambrosini, D.; Sfarra, S. The energy efficiency challenge for a historical building undergone to seismic and energy refurbishment. *Climamed. 2017 Mediterr. Conf. Hvac Hist. Build. Retrofit Mediterr. Area* **2017**, *133*, 231–242. [CrossRef]

17. Muñoz-González, M.C.; León-Rodríguez, L.Á.; Suárez Medina, C.R.; Teeling, C. Hygrothermal performance of worship spaces: Preservation, comfort, and energy consumption. *Sustainability* **2018**, *10*, 3838. [CrossRef]

18. Schibuola, L.; Scarpa, M.; Tambani, C. Innovative technologies for energy retrofit of historic buildings: An experimental validation. *J. Cult. Herit.* **2018**, *30*, 147–154. [CrossRef]

19. Serraino, M.; Lucchi, E. Energy efficiency, heritage conservation, and landscape integration: The case study of the san martino castle in parella (turin, italy). *Energy Procedia* **2017**, *133*, 424–434. [CrossRef]

20. Emmi, G.; Zarrella, A.; De Carli, M.; Moretto, S.; Galgaro, A.; Cultrera, M.; Di Tuccio, M.; Bernardi, A. Ground source heat pump systems in historical buildings: Two italian case studies. *Energy Procedia* **2017**, *133*, 183–194. [CrossRef]

21. Pacchiega, C.; Fausti, P. A study on the energy performance of a ground source heat pump utilized in the refurbishment of an historical building: Comparison of different design options. *Energy Procedia* **2017**, *133*, 349–357. [CrossRef]

22. Pisello, A.L.; Petrozzi, A.; Castaldo, V.L.; Cotana, F. On an innovative integrated technique for energy refurbishment of historical buildings: Thermal-energy, economic and environmental analysis of a case study. *Appl. Energy* **2016**, *162*, 1313–1322. [CrossRef]

23. Ciampi, G.; Rosato, A.; Scorpio, M.; Sibilio, S. Retrofitting solutions for energy saving in a historical building lighting system. *Energy Procedia* **2015**, *78*, 2669–2674. [CrossRef]

24. Lo Basso, G.; Rosa, F.; Astiaso Garcia, D.; Cumo, F. Hybrid systems adoption for lowering historic buildings pfec (primary fossil energy consumption)—A comparative energy analysis. *Renew. Energy* **2018**, *117*, 414–433. [CrossRef]

25. Soflaei, F.; Shokouhian, M.; Zhu, W. Socio-environmental sustainability in traditional courtyard houses of iran and china. *Renew. Sustain. Energy Rev.* **2017**, *69*, 1147–1169. [CrossRef]

26. Vitale, V.; Salerno, G. A numerical prediction of the passive cooling effects on thermal comfort for a historical building in rome. *Energy Build.* **2017**, *157*, 1–10. [CrossRef]

27. Di Turi, S.; García-Pulido, L.J.; Ruggiero, F.; Stefanizzi, P. Recovery of ancient bioclimatic strategies for energy retrofit in historical buildings: The case of the infants' tower in the alhambra. *Energy Procedia* **2017**, *133*, 300–311. [CrossRef]

28. Coillot, M.; El Mankibi, M.; Cantin, R. Heating, ventilating and cooling impacts of double windows on historic buildings in mediterranean area. *Energy Procedia* **2017**, *133*, 28–41. [CrossRef]

29. Rosso, F.; Pisello, A.L.; Castaldo, V.L.; Cotana, F.; Ferrero, M. Smart cool mortar for passive cooling of historical and existing buildings: Experimental analysis and dynamic simulation. *Energy Procedia* **2017**, *134*, 536–544. [CrossRef]

30. Becherini, F.; Lucchi, E.; Gandini, A.; Barrasa, M.C.; Troi, A.; Roberti, F.; Sachini, M.; Di Tuccio, M.C.; Arrieta, L.G.; Pockelé, L.; et al. Characterization and thermal performance evaluation of infrared reflective coatings compatible with historic buildings. *Build. Environ.* **2018**, *134*, 35–46. [CrossRef]

31. Mourão, J.; Gomes, R.; Matias, L.; Niza, S. Combining embodied and operational energy in buildings refurbishment assessment. *Energy Build.* **2019**, *197*, 34–46. [CrossRef]

32. Lidelöw, S.; Örn, T.; Luciani, A.; Rizzo, A. Energy-efficiency measures for heritage buildings: A literature review. *Sustain. Cities Soc.* **2019**, *45*, 231–242. [CrossRef]

33. De Santoli, L. Guidelines on energy efficiency of cultural heritage. *Energy Build.* **2015**, *86*, 534–540. [CrossRef]

34. Litti, G.; Audenaert, A.; Lavagna, M. Life cycle operating energy saving from windows retrofitting in heritage buildings accounting for technical performance decay. *J. Build. Eng.* **2018**, *17*, 135–153. [CrossRef]

35. Blundo, D.S.; Ferrari, A.M.; Fernández del Hoyo, A.; Riccardi, M.P.; García Muiña, F.E. Improving sustainable cultural heritage restoration work through life cycle assessment based model. *J. Cult. Herit.* **2018**, *32*, 221–231. [CrossRef]

36. Loli, A.; Bertolin, C. Towards zero-emission refurbishment of historic buildings: A literature review. *Buildings* **2018**, *8*, 22. [CrossRef]

37. Bertolin, C.; Loli, A. Sustainable interventions in historic buildings: A developing decision making tool. *J. Cult. Herit.* **2018**, *34*, 291–302. [CrossRef]

38. Berg, F.; Flyen, A.-C.; Godbolt, Å.L.; Broström, T. User-driven energy efficiency in historic buildings: A review. *J. Cult. Herit.* **2017**, *28*, 188–195. [CrossRef]

39. Li, Q.; Sun, X.; Chen, C.; Yang, X. Characterizing the household energy consumption in heritage nanjing tulou buildings, china: A comparative field survey study. *Energy Build.* **2012**, *49*, 317–326. [CrossRef]

40. Ben, H.; Steemers, K. Energy Retrofit and Occupant Behaviour in Protected Housing: A case Study of the Brunswick Centre in London. *Energ Build.* **2014**, *80*, 120–130. [CrossRef]

41. Winter, T. Climate change and our heritage of low carbon comfort. *Int. J. Herit. Stud.* **2016**, *22*, 382–394. [CrossRef]

42. Spigliantini, G.; Fabi, V.; Corgnati, S. Towards High Energy Performing Historical Buildings. A Methodology Focused on Operation and Users' Engagement Strategies. *Energy Procedia* **2017**, *134*, 376–385. [CrossRef]

43. Webb, A.L. Energy retrofits in historic and traditional buildings: A review of problems and methods. *Renew. Sustain. Energy Rev.* **2017**, *77*, 748–759. [CrossRef]

44. Hens, H.; Parijs, W.; Deurinck, M. Energy consumption for heating and rebound effects. *Energy Build.* **2010**, *42*, 105–110. [CrossRef]

45. Shove, E. What is wrong with energy efficiency? *Build. Res. Inf.* **2018**, *46*, 779–789. [CrossRef]

46. Guerra Santin, O.; Itard, L.; Visscher, H. The effect of occupancy and building characteristics on energy use for space and water heating in dutch residential stock. *Energy Build.* **2009**, *41*, 1223–1232. [CrossRef]

47. Stern, P.C.; Janda, K.B.; Brown, M.A.; Steg, L.; Vine, E.L.; Lutzenhiser, L. Opportunities and insights for reducing fossil fuel consumption by households and organizations. *Nat. Energy* **2016**, *1*, 16043. [CrossRef]

48. Yarrow, T. Negotiating heritage and energy conservation: An ethnography of domestic renovation. *Hist. Environ. Policy Pract.* **2016**, *7*, 340–351. [CrossRef]

49. Ascione, F.; Bianco, N.; De Masi, R.F.; de'Rossi, F.; Vanoli, G.P. Energy retrofit of an educational building in the ancient center of benevento. Feasibility study of energy savings and respect of the historical value. *Energy Build.* **2015**, *95*, 172–183. [CrossRef]

50. De Santoli, L.; Mancini, F.; Rossetti, S.; Nastasi, B. Energy and system renovation plan for galleria borghese, rome. *Energy Build.* **2016**, *129*, 549–562. [CrossRef]

51. Cassar, M. Sustainable heritage: Challenges and strategies for the twenty-first century. *Apt Bull. J. Preserv. Technol.* **2009**, *40*, 3–11.

52. EFFESUS. *Energy Efficiency in European Historic Urban Districts a Practical Guidance*; Fraunhofer-Center for International Management and Knowledge Economy MOEZ: Leipzig, Germany, 2016.

53. Historic Scotland. *Fabric Improvements for Energy Efficiency*; Historic Scotland: Edinburgh, UK, 2013.

54. Hummelt, K. *Micro-Renewables in the Historic Environment*; Historic Scotland: Edinburgh, UK, 2014.

55. Jenkins, M.; Curtis, R. *Improving Energy Efficiency in Traditional Buildings*; Historic Scotland: Edinburgh, UK, 2014.

56. Sesana, E.; Gagnon, A.; Bertolin, C.; Hughes, J. Adapting cultural heritage to climate change risks: Perspectives of cultural heritage experts in europe. *Geosciences* **2018**, *8*, 305. [CrossRef]

57. Sesana, E.; Bertolin, C.; Loli, A.; Gagnon, A.S.; Hughes, J.; Leissner, J. *Increasing the Resilience of Cultural Heritage to Climate Change Through the Application of a Learning Strategy*; Springer International Publishing: Cham, Switzerland, 2016; pp. 402–423.

58. IPCC. *Climate Change 2014: Mitigation of Climate Change. Contribution of Working Group III to the fifth Assessment Report of the Intergovernmental Panel on Climate Change*; Intergovernmental Panel on Climate Change: Cambridge, UK; New York, NY, USA, 2014.

59. VIOLET. Preserve Traditional Buildings through Energy Reduction. Available online: https://www.interregeurope.eu/violet (accessed on 29 May 2019).

60. Barca, D.; Comite, V.; Belfiore, C.M.; Bonazza, A.; La Russa, M.F.; Ruffolo, S.A.; Crisci, G.M.; Pezzino, A.; Sabbioni, C. Impact of air pollution in deterioration of carbonate building materials in italian urban environments. *Appl. Geochem.* **2014**, *48*, 122–131. [CrossRef]

61. Historic Scotland. *A Climate Change Action Plan for Historic Scotland 2012–2017*; Historic Scotland: Edinburgh, UK, 2012.

62. Gelowitz, M.D.C.; McArthur, J.J. Comparison of type III environmental product declarations for construction products: Material sourcing and harmonization evaluation. *J. Clean Prod.* **2017**, *157*, 125–133. [CrossRef]
63. González Longo, C. Can Architectural Conservation Become Mainstream? In Proceedings of the 18th Scientific Symposium (ICOMOS), Florence, Italy, 10–14 November 2014.
64. Sabbioni, C.; Cassar, M.; Brimblecombe, P.; Lefevre, R.A. *Vulnerability of Cultural Heritage to Climate Change*; European and Mediterranean Major Hazards Agreement (EUR-OPA), Council of Europe: Strasbourg, France, 2008; pp. 1–24.
65. Cantin, R.; Burgholzer, J.; Guarracino, G.; Moujalled, B.; Tamelikecht, S.; Royet, B.G. Field assessment of thermal behaviour of historical dwellings in france. *Build. Environ.* **2010**, *45*, 473–484. [CrossRef]

Article

Categorization of South Tyrolean Built Heritage with Consideration of the Impact of Climate

Lingjun Hao [1,2,*], Daniel Herrera-Avellanosa [2], Claudio Del Pero [1] and Alexandra Troi [2]

[1] Department of Architecture, Built Environment and Construction Engineering, Politecnico di Milano, 20133 Milano, Italy; Claudio.delPero@polimi.it
[2] Institute for Renewable Energy, Eurac Research, 39100 Bolzano, Italy; Daniel.Herrera@eurac.edu (D.H.-A.); Alexandra.Troi@eurac.edu (A.T.)
* Correspondence: Lingjun.Hao@polimi.it;

Received: 31 October 2019; Accepted: 4 December 2019; Published: 9 December 2019

Abstract: Climate change imposes great challenges on the built heritage sector by increasing the risks of energy inefficiency, indoor overheating, and moisture-related damage to the envelope. Therefore, it is urgent to assess these risks and plan adaptation strategies for historic buildings. These activities must be based on a strong knowledge of the main building categories. Moreover, before adapting a historic building to future climate, it is necessary to understand how the past climate influenced its design, construction, and eventual categories. This knowledge will help when estimating the implication of climate change on historic buildings. This study aims at identifying building categories, which will be the basis for further risk assessment and adaptation plans, while at the same time analyzing the historical interaction between climate and human dwelling. The results show some correlations between building categories and climate. Therefore, it is necessary to use different archetypes to represent the typical buildings in different climate zones. Moreover, these correlations imply a need to investigate the capability of the climate-responsive features in future climate scenarios and to explore possible further risks and adaptation strategies.

Keywords: built heritage; categorization; climate change adaptation

---

1. Introduction

Building categorization allows dividing the building stock into several homogeneous building groups according to certain key features such as construction period, building volume, material, etc. Archetypes or reference buildings could be selected from each building group to represent the most significant categories/typologies of the building stock. Obviously, this is only possible within certain assumptions and limits. Yet, these simplifications of the building stock are necessary for policy development and any other activity that aims at addressing the whole built heritage stock. By offering such archetypes, building categorization supports a bottom-up analysis of the building stock that allows an assessment of the energy consumption and potential conservation threats to a large building stock [1–3].

In the case of historic buildings, however, local influences on building typology due to factors like the evolution of the economic structure, population concentration, and diffusion will challenge the generalizing approach of categorization [4]. The combination of all these factors results in an intricate history of design, construction, and renovation process of the buildings, which makes each historic building unique and hard to be grouped. However, climate change mitigation and adaptation activities require a certain generalization of the built stock.

The severity and impact of climate change were rigorously assessed in scientific literature. According to IPCC's (Intergovernmental Panel on Climate Change) Fifth Assessment report [5], the increase in global surface temperature by the end of the 21st century is expected to exceed 2.6–4.8 °C compared to 1986–2005 in the most pessimistic scenario. Together with this temperature increase, extreme climate events are expected to occur more frequently [5,6]. In South Tyrol, climate change is clearly apparent in the increasing temperature and changed precipitation pattern. For instance, there would be more tropical nights (Nights during which the temperature remains above 20°C) in summer and more precipitation in winter. Moreover, heat waves and extreme rain events would be more frequent [7]. It is also urgent to decrease the greenhouse gas emission in the built heritage sector to mitigate climate change. One of the barriers to climate change mitigation is the incompatible retrofit solutions [8]. Climate change and incompatible solutions impose great challenges on the built heritage sector by increasing the risks of energy inefficiency, indoor overheating, and moisture-related damage to the envelope [9]. To precisely identify the effect of climate change on the performance of retrofitted historic building, a three-year research project is being conducted, which includes four steps: (1) the identification of building categories and reference buildings (partly presented here), (2) the identification and assessment of present retrofit solutions, (3) the assessment of the combined impact of climate change and retrofit solutions, and (4) based on the results of the previous steps, suggestions for adaptation measures that are compatible with present and future weather. In this paper, the methodology to categorize historic buildings is presented. This will be the basis for further climate mitigation and adaptation studies.

Among the influencing factors, culture background, social customs, and most importantly the climate should be emphasized. Climate variability can impact culture, landscape, and human settlement [10–12]. Moreover, many studies confirmed the relationship between building characteristics and the local climate. In fact, the morphology, the position and size of windows, the wall material, etc. of historic buildings present climate-responsive features [13–16]. In Alpine regions, a wide range of landscapes and buildings evolved in the process of inhabitants' adaptation to local climate. They are a constitutive and essential part of the Alpine identity, sharing similarity in reflecting Alpine living. Building settlement form, construction technique, and other morphological or technical characteristics display the logic of climate adaptation [4,17]. For instance, the masonry is constructed with two external stone layer fillings with aggregates bonded with earth mortar and lime mortar to resist harsh external conditions; the compact volumes limit the thermal dispersion; the size and position of the windows are designed to minimize heat losses; the unoccupied attics reduce the heat loss through the roof thanks to the storage of hay or other fodder [18,19].

South Tyrol is a typical Alpine region in the north of Italy. It is characterized by its mountainous topography and diverse climatic conditions. Consequently, it offers a good scenario for the analysis of the relationship between climate and building typology evolution.

In summary, climate may have formed the typology of historic buildings in South Tyrol to some extent. Considering severe climate change in the future, historic buildings that were designed, constructed, and renovated according to climatic conditions in the past may be vulnerable to new threats, which will affect their conservation or performance in terms of indoor comfort and energy consumption. Conducting a categorization with a special focus on climate allows analyzing the historical interaction between climate and human dwelling activities and, accordingly, verifying the possible effects of future climate on historic buildings. Furthermore, archetypes representing the main categories could facilitate assessing the performance of the built heritage stock and planning the adaptation strategies in changing climate context. This study focuses on listed historic buildings [20], since they have the priority to be conserved and retrofitted, and specifically on residential buildings, as they represent the largest portion of the listed stock in South Tyrol and most parts of Europe [21,22].

2. Building Categorization: A Critical Review of Existing Methodologies

Building categories enable grouping different buildings that have similar or comparable features with the scope of being representative. The number of descriptive features depends on the number of target buildings, available building inventory, etc. There are no standardized characteristics; requirements and characteristics are selected for the purpose of the categorization.

In recent studies, one of the most common categorization targets was to support the assessment of the energy consumption or emission of the building stock (Table 1), i.e., to establish a stock energy/emission model. In that case, archetypes are created representing each category before scaling their energy use according to individual impacts to model the energy use of the entire stock. [23]. In the literature, energy use-related factors such as geometrical and thermal–physical properties of the building, the heating and cooling system, the climate zone of the building, etc. are used in categorization [2,24,25]. However, selecting all the variables that are significant for building energy performance is not feasible due to the data availability and the complexity of the energy model. Famuyibo et al. [2] attempted to define the key variables of buildings based on their impact on energy use (Table 1). Through multiple linear regression analysis, typical weekly occupancy pattern (heating season) (low/medium/high), internal temperature (°C), immersion heater weekly frequency, and air change rate (ac/h) were selected from existing inventories because they are significant variables that influence the total energy use. However, it was found that, due to the limitations of the dataset (lack of data such as occupancy behavior), more than 60% of the energy use variation could not be explained by the model. Moreover, the first three of the significant variables were excluded since occupant-related variables were standardized in the operation of a reference building.

In the case of historic buildings, categorization aims to support not only energy performance assessment but also risk mitigation and the identification of retrofit solutions (Table 1). In some cases, it is used as a process to analyze historic buildings through identifying the vernacular characteristics, cataloging the materials in the different construction periods, etc. [1,26]. Similar to non-historic building categorization, geometrical characteristics such as floor area and number of stories are adopted due to the general availability and their close relation to building energy performance. Thermal and hygrometric features such as construction materials are important for the preservation of heritage and the selection of retrofit solutions; therefore, they are generally used in categorization. In addition to that, the protection degree or other legislative requirements are included in some cases to present the historic significance or renovation limits of the buildings [27,28]. Construction period is selected because it reveals further information about building typology, construction materials, building equipment, etc. [3,25,27], thereby implying an analysis of the social, legislative, and technical impacts on building typology. Moreover, features on the settlement level could present the rooting of building stock. Montalbán Pozas and Neila González [1] suggested that categories of historic buildings should consider the sociocultural, economic, and historical contexts. They identified the building categories in a historic stock according to key features on four levels: territory, urban planning, architecture, and construction process, where features like the width and orientation of the streets, typical parceling of the blocks, etc. help interpreting the development and habitability problems of the stock.

There remains the problem of lacking data [23]. Since historic buildings have a complex history of construction and repairs, survey work covering the whole building stock is still infrequent. To avoid using deficient data, qualitative approaches are conducted in studies, such as expert evaluations, literature reviews, and on-site surveys [1,3,25,29]. For instance, due to the lack of adequate statistical data, the categorization of Hungarian stock was based on expert judgements [3]. A qualitative study could help understanding the building typology from a genealogy point of view, focusing on how the typologies evolved [30]. It helps linking the typology with its historic context.

Once the key features are selected, the category structure could be defined. There are two main category structures: flow structure and matrix structure, as shown in Figure 1. The category process of a flow structure successively divides the whole building stock according to selected features. The matrix structure is formed with two main key features. For instance, in the TABULA (Typology Approach for Building Stock Energy Assessment) project [31], building types (single-family houses, terraced houses, and blocks) and construction periods were selected as the two main features. Both structures have strengths and weaknesses. For the flow structure, it could include enough key features to establish detailed building categories, but the key features and intervals should be carefully determined since too many categories could be generated and some categories may not be representative. For the matrix structure, only two key features are involved in categorization; therefore, other features should be carefully added into the description of archetypes, without influencing the category results.

Table 1. Variables found in literature for building categorization.

Study	Country	Specific to Historic Buildings	Aim of the Categorization	Key Features Used to Describe the Building Categorization
[2]	Ireland	No	To support evidence-based energy and emissions policy	Air change rate, wall, roof, floor, and window U-values, dwelling type (number/area of external walls), heating and domestic hot water system, floor area
[24]	Italy	No	To assess the energy requirements of the residential stock	Construction period, geometrical properties, thermo-physical properties, heating system
[32]	Italy	No	To understand the energy performance of the building stock	Construction period
[33]	Greece	Partly	To plan and promote new energy renovation scenarios	Construction period, building use, number of floors, material
[3]	Hungary	Partly	To assess the vulnerability of building stock to increasing wind	Construction period, construction type, roof configuration, number of stories, building surroundings, materials of the building envelope
[28]	Italy	Yes	To develop energy analyses for regulation and financial strategies	Level of protection, building volume, organization of indoor spaces and adjacent constructions, thermal and hygrometric properties of envelope components
[27]	Spain	Yes	To support energy retrofit of the building stock	Main use, number of facades, year of construction, protection degree, volume
[34,35]	Sweden	Yes	To assess and select energy efficiency interventions	Climate zone, type of building, use, size, age of construction, aggregation with adjacent constructions, heating system
[26]	Malaysia	Yes	To improve knowledge and preservation of the built heritage	Demography, ownership, type of settlement, historical background, geographic location, landscape features, communication, accessibility, and surroundings
[25]	Eastern Europe	Yes	To assess the energy requirements and saving potential of the residential stock	Country, construction year, building size
[29]	Portugal	Yes	To support risk mitigation at urban scale	Building size, configuration, and volume, number of floors, distribution systems, building materials, construction period
[1]	Spain	Yes	To provide guidelines for the analysis of historic buildings	Features on four scales: territory, urban planning, architecture, and construction

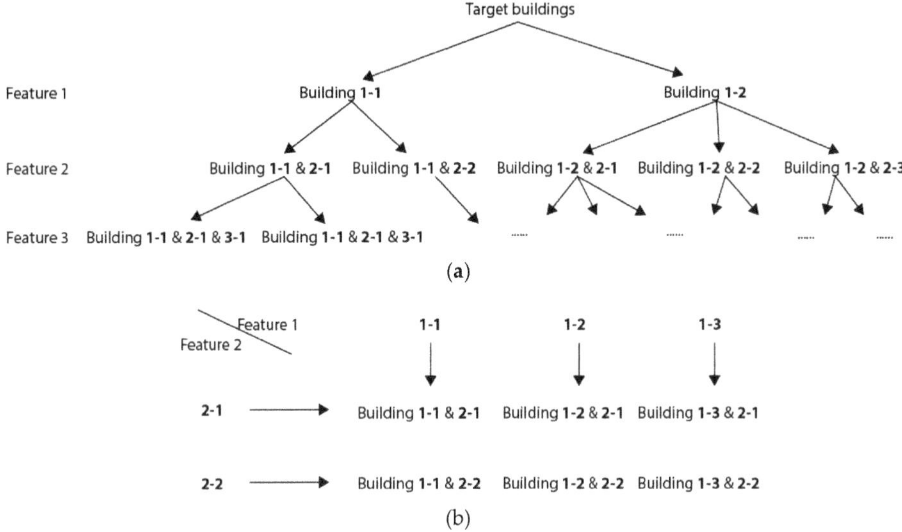

Figure 1. Category structure: (**a**) flow structure; (**b**) matrix structure.

3. Proposed Methodology

The methodology proposed in this paper was developed to prepare building categories for further risk assessment and adaptation planning, while permitting the possibility to analyze the relationship between climate and building categories which would provide knowledge support for further studies.

In order to identify the relationship between building categories and climate, the climate of South Tyrol was firstly analyzed and subdivided into homogeneous zones (Figure 2, step 1). In each climate zone, building samples were randomly extracted from the building stock (Figure 2, step 2a). Probability sampling was adopted in this study to ensure the representativeness of the sample despite the limited research resources.

At the same time, key features were defined according to the aim of the categorization (Figure 2, step 2b) through a literature review including categorization studies and studies on South Tyrolean residential buildings. Experts were consulted on whether the key features were representative and feasible to be used in this study. The criteria to select the expert panel were as follows: people who share an interest in the research project, and who have the knowledge of South Tyrolean historic buildings or have the experience of building categorization. In this study, the expert panel included three researchers based in South Tyrol with an expertise on energy renovation of historic buildings, as well as a local architect specialized in the conservation and adaptation of South Tyrolean heritage. Then, the defined representative features were collected for the building samples (Figure 2, step 3a), from available building inventories and the literature (step 3b).

After the dataset of key features was established, it was used in a flow structure to categorize the building samples (Figure 2, step 4). Eventually, the key features of the categories were statistically analyzed and compared among different climate zones (step 5a, Figure 2).

The results were interpreted with a qualitative study of South Tyrolean historic buildings, climate conditions, historic and social–economic events which influenced building customs, etc. (step 5b). By tracing the development of historic buildings, the relationship between climate and building categories was analyzed (step 6, Figure 2).

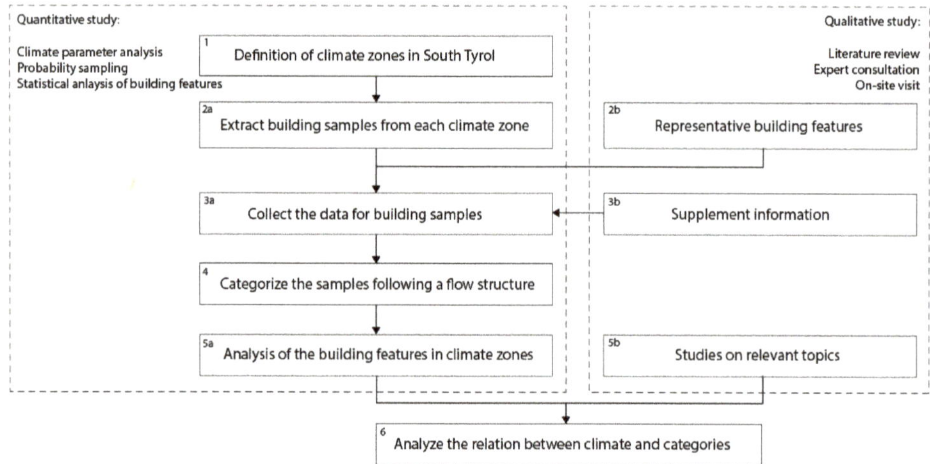

Figure 2. The methodology for categorization.

4. Categorization of the Historic Building Stock in South Tyrol

4.1. Climate Zone of South Tyrol

The whole region of South Tyrol covers 7400 km^2 with altitudes ranging from 190 m to more than 3000 m (Figure 3). The surface area below 1000 m above sea level (a.s.l.) is 14.1% of the total area, while the surface area over 1500 m a.s.l. represents 64.4% of the total area [36]. Due to the mountainous topography, diverse climate conditions exist. To analyze and subdivide the climate, climate data of different locations in South Tyrol are required. In this paper, climate data used were from (1) Provincia Autonoma di Bolzano Alto Adige (including data of 30 representative weather stations), and (2) results of the 3PClim project [37].

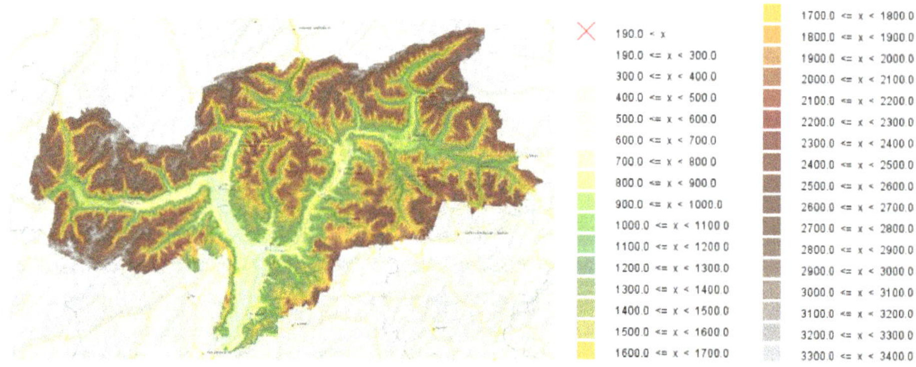

Figure 3. Elevation of South Tyrol (extracted from digital terrain model, http://geokatalog.buergernetz.bz.it/geokatalog/#!).

In this study, sub-climate types were defined according to criteria introduced below, which describe the similarities and distinctions in climate patterns. The climate zones were generated based on the results of the 3PClim project, where geostatistical interpolation methods were applied with the aid of programming software and geographical information system software [37].

The descriptive criteria were defined with the consideration of Köppen climate classification [38], which is a widely used climate classification system. The main parameters used in Köppen climate classification are annual and monthly sums of precipitation, and annual and monthly mean temperature. The fundamental scheme of climate classification includes five major climate types (tropical, dry, temperate, continental and polar) covering the whole global climate. According to Köppen climate classification, the weather stations found in South Tyrol would fall into four different climate zones: Cfa, Cfb, Dfb, and Dfc. The differences between the four climate zones are shown in Table 2, and they are all temperature factors. However, the precipitation varies largely in South Tyrol from a regional point of view (Figure 4). Since precipitation has a significant impact on a building's hygrothermal performance, it is necessary to include precipitation in the climate zone definition in this study.

Table 2. Climate differences among four climate zones defined by Köppen climate classification. T—temperature.

	Cfa	Cfb	Dfb	Dfc
Average T of the coldest month	0 °C–18 °C	0 °C–18 °C	≤0 °C	≤0 °C
Average T of the warmest month	≥22 °C	<22 °C	-	-
No. of months with average month T ≥10 °C	-	≥4	≥4	<4

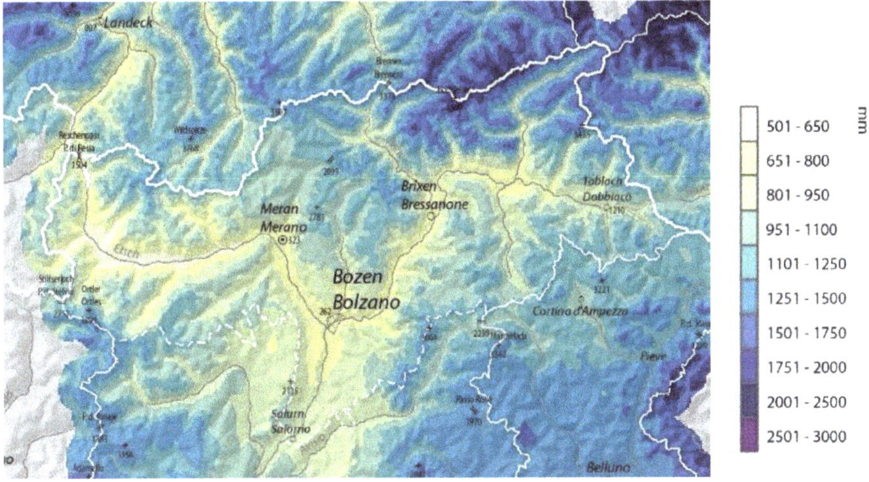

Figure 4. Mean annual total precipitation of South Tyrol (reference period: 1981–2010, http://www. 3pclim.eu/).

In this study, the average temperature of the coldest month was used to divide the relatively warm zones from the relatively cold zones. To emphasize the impact of precipitation on climate classification, the median amount of precipitation of 30 representative weather stations was introduced as a criterion to differentiate between relatively dry and relatively wet zones. According to these two criteria, four sub-climate zones were defined (Table 3).

As shown in Figure 5, Zone I lays at the southern part of South Tyrol, covering regions with an altitude below 800 m. Zone I covers Val d'Adige that stretches from Salorno northward to Merano, and runs westward along Val Venosta to Naturno. In the east, it covers a narrow strip of low land along Valle Isarco. Zone I also includes the southern part of Val Sarentino that has relatively low altitude. The climate of Zone I is characterized by relatively warm temperatures and less precipitation. Compared to Zone I, Zones II and III have lower temperatures generally. Zone II distributes mainly in two parts: (a) the western part of South Tyrol, which includes Val Venosta and its side valleys such as Val Senales, Val di Trafoi, Val Martello, and Val d'Ultimo below 1300 m in elevation, and (b) the eastern part comprising the districts of Val d'Adige and Valle Isarco, where the altitude is around

600 m–1300 m, as well as Val Pusteria and its side valleys. The climate of Zones II and III differs in precipitation (Zone II has less precipitation). Zone III includes the vast highland in central and eastern South Tyrol. A fourth climate zone exists but is not included in this study due to its limited presence in the region.

Table 3. Climate differences among climate zones in this study.

	Zone I	Zone IV	Zone II	Zone III
Average T of coldest month	0 °C–18 °C	0 °C–18 °C	≤0 °C	≤0 °C
Average annual precipitation	≤825.2 mm	>825.2 mm	≤825.2 mm	>825.2 mm
Feature	Relatively warm and dry	Relatively warm and wet	Relatively cold and dry	Relatively cold and wet

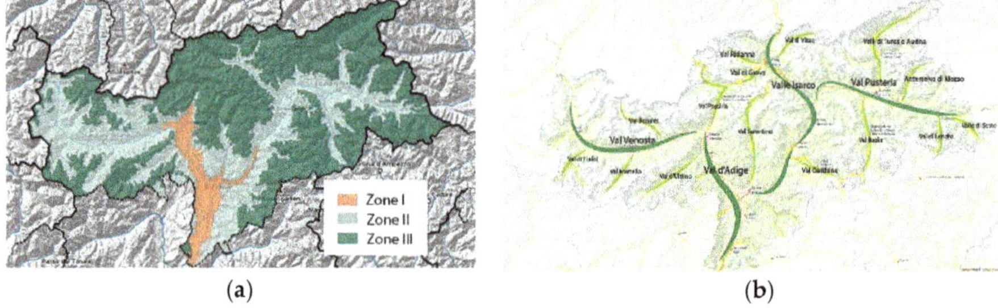

(a) (b)

Figure 5. (a) Climate zones in South Tyrol; (b) main valleys in South Tyrol.

4.2. Historic Residential Buildings in South Tyrol

According to the 2017 population census of South Tyrol [39], the residential stock is composed of 225,483 buildings in total [39], 34,160 of which were built before 1919 and 14,840 of which were built during 1919–1945. These two parts comprise 22% of the total stock (Figure 6). Only 10.4% of this stock was retrofitted in the past 10 years (Figure 6). While energy retrofit represents an opportunity to reduce a building's operational energy and CO_2 emissions, forecasted climate change might impose great risks on the hygrothermal performance of building constructions after retrofitting. For this reason, there is an urgent need to analyze the relationship between building categories and climate, and to prepare archetypes for further performance assessment.

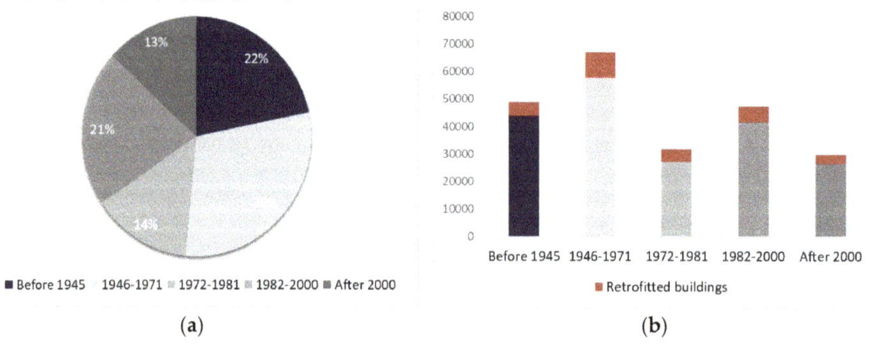

(a) (b)

Figure 6. (a) Residential buildings by construction period; (b) residential buildings retrofitted during last 10 years by construction period [39].

Among the large residential stock, 4537 residential buildings in three categories (rural buildings, urban buildings, and nobility buildings, Figure 7) are listed as historic buildings under protection. Since rural farmhouses form the outstanding landscape of the Alpine space, they were selected to be studied under the category of rural buildings (Figure 7). In urban buildings, the trade–residential nucleus, the Portici house (Figure 7), was studied because it is the most important urban residence in the culture, social, and economic centers of the cities in South Tyrol. It appears in Merano, Bolzano, Egna, Bressanone, Vipiteno, and Glorenza. For rural farmhouses and Portici houses in each climate zone, building samples were randomly extracted ensuring a confidence interval lower than 15% and a confidence level of 95%, as shown in Table 4.

Even though the application of the proposed methodology is on listed historic buildings, it could be used for any historic building stock.

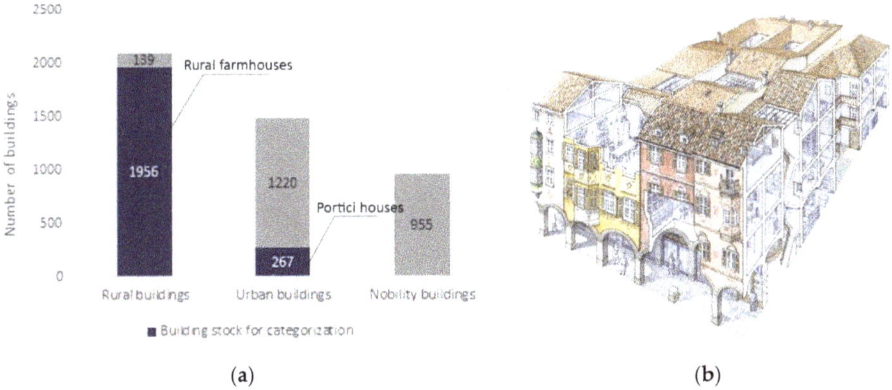

Figure 7. (**a**) Listed residential buildings in South Tyrol, and the building stock for categorization (data source: http://www.provinz.bz.it/kunst-kultur/denkmalpflege/monumentbrowser-suche.asp); (**b**) scheme of a Portici house (© Antonio Monteverdi, http://www.antoniomonteverdi.com/sito/?page_id=1228).

Table 4. Information about building samples.

Climate Zone	Residence Type	No. of Buildings	Sample Size	Sample Size (%)	Confidence Level	Confidence Interval
I	Rural farmhouses	628	90	14.3%	95%	10%
	Portici houses	148	35	23.6%	95%	15%
II	Rural farmhouses	748	90	12.0%	95%	10%
	Portici houses	119	35	29.4%	95%	15%
III	Rural farmhouses	580	85	14.7%	95%	10%

4.3. Key Feature Selection

Our literature review showed that the key features used for any categorization should be selected according to the targets of the categorization. Therefore, features that are performance-related and potentially climate-responsive were selected to construct the flow structure for categorization. To better reflect the influence of climate, geographic condition, and historical context, key features were selected in three scale levels: settlement scale, building scale, and element scale.

On the settlement scale, key features included the compactness of the settlement and the number of the adjacent walls of buildings. Here, "compactness" describes the concentration level of the buildings; a "compact" type means that most of the buildings in this settlement are surrounded by close obstacles, while a "sparse" type means that most of the buildings are exposed to wind and rain without close obstacles. Close obstacles are defined as obstacles with a maximum distance of 25 m, which refers to the obstruction factor of 0.4 in EN 15927-3 [40]. The number of adjacent walls expresses the density of the settlement layout. The compactness of the settlement influences a building's resilience to extreme climates, while the number of the adjacent walls influences the energy use of the building and the indoor thermal comfort.

On the building scale, the typical Alpine building forms were considered. Geometrical and thermophysical-related features including roof projection area, floor number, window-to-wall ratio, and construction material were collected. Data were taken from existing GIS (Geographic information system) maps from GeoKatalog of Province Bolzano [41], external inspections, and photo evaluations. Geometrical features may result in different energy performance and thermal comfort according to the literature review, and construction materials present different behaviors in terms of moisture dynamics. The building layout, which indicates the distribution of functional space, was studied from the literature as supplementary information. The layout of residence space, farm space, and commercial space influences the heating setpoint, i.e., the heating schedule of the building spaces; therefore, it affects energy consumption.

Valuable building elements, which have historic, cultural, natural, morphological, and aesthetic value, were summarized from the literature because they are the essence of historic buildings and a crucial factor in retrofit decision-making. Any retrofit solution should be compatible with the heritage elements. Therefore, they influence the performance of retrofitted buildings indirectly.

4.4. Building Categories

To define a reasonable number of building categories, settlement compactness, construction material, and the number of floors were used to construct the flow structure of the categorization, while other features were used as supplementary information. Therefore, 12 building categories, representing 81.6% of the building samples, are defined for further study (Table 5). All the key features are compared among different climate zones in the subsequent sections.

Table 5. Building categories of historic residential buildings.

Climate Zone	Residence Type	Settlement Type	Construction Material	Number of Floors	Category Code [1]	Building Layout	Adjacent Walls	Window-to-Wall Ratio (%)	Roof Projection Area (m²)
I	Rural farmhouse [2] (R)	Compact (C)	Masonry and wood (attic only) (MW)	3 (3f)	I-R-C-MW-3f	Paarhof [3]	1	0.17–0.2	340
				2 (2f)	I-R-C-MW-2f				
		Scattered (S)	Masonry and wood (attic only) (MW)	3 (3f)	I-R-S-MW-3f		0		
	Portici house [4] (P)	Compact (C)	Masonry (M)	4 (4f)	I-P-C-M-4f	Portici house	2	0.21–0.4	447.6
				3 (3f)	I-P-C-M-3f				
II	Rural farmhouse (R)	Compact (C)	Masonry and wood (attic only) (MW)	3 (3f)	II-R-C-MW-3f	Paarhof	1	0.12–0.19	304
				2 (2f)	II-R-C-MW-2f				
		Scattered (S)	Masonry and wood (attic only) (MW)	2 (2f)	II-R-S-MW-2f		0		
	Portici house (P)	Compact (C)	Masonry (M)	4 (4f)	II-P-C-M-4f	Portici house	2	0.15–0.35	360.1
				3 (3f)	II-P-C-M-3f				
III	Rural farmhouse (R)	Scattered (S)	Masonry and wood (attic only) (MW)	2 (2f)	III-R-S-MW-2f	Paarhof	0	0.07–0.14	270
			Masonry and wood (first floor and attic) (MWW)	2 (2f)	III-R-S-MWW-2f				

[1] The category code is formed with the initials showed in brackets in the columns on the left; [2] Elements worthy of preservation: façade decoration (e.g., fresco painting, stucco), internal fitting (e.g., carved ceiling, wood-panelled wall), historic windows, wood construction (e.g., Blockbau, Ständerbohlenbau), historic roof, vault construction, etc.; [3] By the term of Paarhof, a farm layout is described where the dwelling building and the farm building stand independently. More information could be found in 5.2.1; [4] Elements worthy of preservation: façade decoration (e.g., fresco painting, stucco), arcades, bay windows, wrought-iron rails, stone stairs, etc.

5. Results and Discussion: The Impact of Climate on the Development of Dwelling in the Alps

In this section, the differences in the key features of historic buildings in three climate zones are presented, as a result of the quantitative study. To interpret and discuss these differences, we made use of the qualitative information resulting from the study of building history. Discussions focus on the differences that were historically influenced by climate to explore the possible role of climate in shaping the building categories.

5.1. Settlement Level

5.1.1. Rural Farmhouse

- Description of quantitative results

According to the sampling survey (Table 6), in climate Zone I, 75.3% of the buildings are in compact settlements, while 44.9% of the buildings are semi-detached (one adjacent wall) and 42.7% are detached. In climate Zone II, the settlements are less concentrated, whereby 55.1% of the buildings are in compact settlements while the others are in sparse settlements. More than 66% of the farmhouses are detached. Climate Zone III has 67.7% of farmhouses in sparse settlements, whereas more than 90% of the farmhouses are detached buildings.

Table 6. Rural settlement comparison in three climate zones.

Climate Zone	Zone I	Zone II		Zone III
Settlement type	Compact settlements (75.3%)	Compact + sparse settlements (55.1% + 44.9%)		Sparse settlements (66.7%)
Typical Diagram	Termeno	Chienes	Val di Vizze	Selva dei molini
Adjacent walls	0 (42.7%) 1 (44.9%)	0 (66.3%)	1 (28.1%)	0 (91%) 1 (9%)
Picture	Merano [1]	Silandro [2]	Silandro [3]	Ortisei [4]

[1] https://pxhere.com/de/photo/1095092; [2] https://www.iha.com.de/ferienwohnungen-schlanders-silandro/ig!/; [3] Office of Architectural and Artistic Heritage, Autonomous Province of Bolzano - Alto Adige, http://www.provinz.bz.it/kunst-kultur/denkmalpflege/monumentbrowser-suche.asp?status=detail&id=17282; [4] Wolfgang Moroder, "Der Bauernhof Peza in St. Ulrich in Gröden", https://commons.wikimedia.org/wiki/File:Peza_Sacun_Urtijei_dinsta.jpg, 2016.

- Discussion with consideration of qualitative results

The concentration of the buildings and the density of the settlements decrease from climate Zone I to Zone III. This could be due to the interaction within social development, environment availability, and climate diversity. Climate and nature resources are important driving factors for settlement development, especially before modern history, when humans had less resilience against environmental changes. In the north and south of the Alps, periods of warm climate were observed to coincide with land-use expansion and increases in population, while the deteriorated climate was accompanied by land abandonment and reforestation [42]. Through influencing the land use, productivity in agriculture and pasture and the climate variety shape the socio-economic structure, which leads to the concentration of settlement in the long term. The climate in Zone I is more suitable for economic activities compared to that in Zones II and III, which explains the compactness of settlement to some extent. Socio-economic activities and other anthropogenic processes that influence settlements could be seen as reactions to the climate variety [43].

However, climate is not the only factor that determined the form of settlements. Driving factors from the human culture system brought profound changes to settlements of South Tyrol. Most current settlements emerged or consolidated during the Roman dominion, before which Alpine regions were controlled by self-sufficient tribes [44,45]. The stage stations, garrisons, and markets arranged along the Roman road became the first nuclei in Alpine cities [45]. Furthermore, the distribution of different people may have initiated the differences in settlement form and function. Two distinctive administrative structures, the Romanzo or Rhaetian-Romanzo system and Germanic system [45], resulted in two settlement forms. In the Romanzo system, new settlements emerge in a concentrated style to save space and maintain sufficient land for the whole community. In the Germanic system, the landlords manage the settlement forms and entrust the farms to the peasantry in sparsely populated areas. The settlements are scattered away from each other. In summary, the development of South Tyrolean settlements and the compactness of the settlements are the results of a mutual adaptation between the climate and culture system.

5.1.2. Portici House

- Description of quantitative results

According to the sample survey (Table 7), all the settlements of the Portici house are in compact form, and most of the buildings have two adjacent walls. When comparing the size of the settlements in climate Zones I and II, the dimension of settlements is generally larger in Zone I (notice the length of the Portici district in Table 7). Furthermore, although all settlements have a high density, there is a difference in the aspect ratio (distance to height ratio, D/H) of the main street in different climate zones (Figure 8).

- Discussion with consideration of qualitative results

The compact form of the Portici settlements is mainly attributed to the requirement of trading activities. In the late Middle ages, a significant climate warming [46] and political consolidation integrated the Alpine region into the urban expansion progress in Europe. The trading and market activities on trans-Alpine routes pushed the development of urban residences in South Tyrol. During the 11th to 13th century, several villages were chartered as cities and granted market rights, which promoted the prosperity of the city and developed local markets and crafts. In Bolzano, bishops of Trent expropriated a piece of land and divided it into parcels during 1022–1055; these trading–residential parcels are called "Laubengasse" or "Via Portici" [47] (Figure 9). These buildings were highly compact to save public land, and they had a uniform building structure, ordered ridge heights, and a controlled alignment line. Along the continuous façade, there are arcades covering the walkway on the ground floor which form an extension space for trade activities. This is a typical Romanesque building model spreading from the southeast of Bavaria to Tyrol, and westward to eastern Switzerland and southern France [48]. Around the trading district, walls, moats, and towers were built to protect the city. Two gates for the trading routes opened at the west and east ends of the "Via Portici". Outside of the walls, there are farmlands and some farmhouses. Due to wars or traffic reasons, the walls were generally demolished later, and, with city expansion, "new" buildings were built surrounding the Portici settlements (Figure 9).

Table 7. "Portici" settlement comparison in different climate zones [47,49].

Climate Zone	Zone I			Zone II		
"Portici" settlements						
	Bolzano	Merano	Egna	Bressanone	Vipiteno	Gloranza
Street width	4.8–5.8	4.5–6.8	5.2–7.1	6.2–7.2	7.3–8.3	5.2–6.6
No. of Floors	4 (72.7%), 5 (27.3%)	3 (85.7%), 4 (14.3%)	3 (74%), 4 (24%)	4 (71.4%)	4 (74%)	3 (63%), 2 (37%)
Axis	East–west	East–west	East–west	East–west	North–south	East–west
Length	~300 m	~400 m	~250 m	~200 m	~170 m	~100 m
Picture						
	Portici 65–67 [1]	Portici 110–120 [2]	Via Andreas Hofer 14 [3]	Via Portici Maggiori 6 [4]	Gasthof Goldenen Adler [5]	Via Portici 7 [6]

[1] Office of Architectural and Artistic Heritage, Autonomous Province of Bolzano - Alto Adige, http://www.provincia.bz.it/arte-cultura/beni-culturali/monumentbrowser-ricerca.asp?status=detail&id=13870; [2] Office of Architectural and Artistic Heritage, Autonomous Province of Bolzano - Alto Adige, http://www.provincia.bz.it/arte-cultura/beni-culturali/monumentbrowser-ricerca.asp?status=detail&id=15965; [3] Office of Architectural and Artistic Heritage, Autonomous Province of Bolzano - Alto Adige, http://www.provincia.bz.it/arte-cultura/beni-culturali/monumentbrowser-ricerca.asp?status=detail&id=16308; [4] Office of Architectural and Artistic Heritage, Autonomous Province of Bolzano - Alto Adige, http://www.provincia.bz.it/arte-cultura/beni-culturali/monumentbrowser-ricerca.asp?status=detail&id=14140; [5] Piergiuliano Chesi, https://commons.wikimedia.org/wiki/File:Vipiteno_Gasthof_Goldenen_Adler.JPG, 2010; [6] Office of Architectural and Artistic Heritage, Autonomous Province of Bolzano - Alto Adige, http://www.provinz.bz.it/kunst-kultur/denkmalpflege/monumentbrowser-suche.asp?status=detail&id=14934.

The east–west axis of the Portici settlement could help in blocking the wind in winter and creating a comfortable local microclimate on the street. Furthermore, the low aspect ratio found in climate Zone I (Figure 9) helps in shading the street in summer, while the higher aspect ratio in climate Zone II permits more sunshine for the buildings. Further studies should be conducted on the impact of the aspect ratio on energy use. Differences were found in the length of the Portici settlements, while there is no clear evidence that climate difference led to this phenomenon. It may be related to the trading scale and land price of the city.

Figure 8. The aspect ratio of Portici houses, left: Bolzano (climate Zone I), right: Vipiteno (climate Zone II).

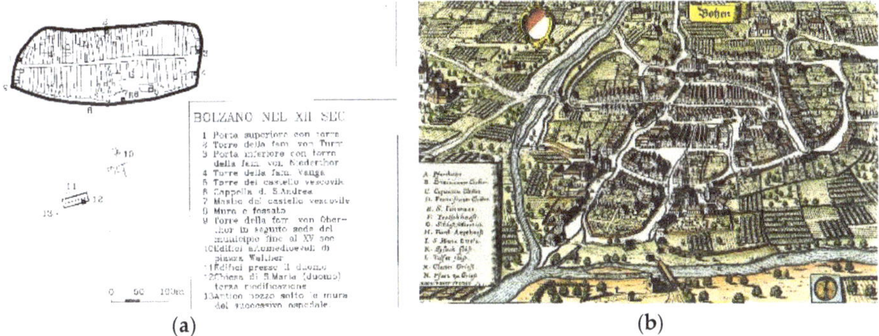

(a) (b)

Figure 9. (**a**) Detailed plan of Bolzano at the end of the 12th century; (**b**) Bolzano in 1645, copper engraving by Matthaeus Merian [47].

5.2. Building Level

5.2.1. Rural Farmhouse

- Description of quantitative results

According to the sample survey, there is a significant difference in the material–use ratio. Masonry buildings are dominant in climate Zone I (Table 8), where about 77.5% of rural farmhouses are constructed in masonry and the rest of the buildings are constructed with wooden attics. In climate Zone II, the use of wood increases. About 46.1% of the buildings are built in masonry, and 39.3% are constructed in masonry with wooden attics. Furthermore, 15.6% of masonry buildings have wooden floors. In climate Zone III, the wood ratio increases further compared to the other climate zones. Pure masonry buildings account for 26.7% while 26.7% of the masonry buildings have wooden attics, and

46.7% of the masonry buildings have wooden floors and attics. The window-to-wall ratio (W-to-W) decreases from Zone I to III. The dimension of the rural buildings in the three climate zones also varies (Table 8). The average area of roof projection decreases from climate Zone I to III with average numbers of 340 m^2 to 304 m^2 and 270 m^2, and the typical number of floors above ground decreases from three to two.

Table 8. Building features of rural farmhouses in three climate zones.

Climate Zone	Zone I		Zone II			Zone III		
Material	M	MW	M	MW	MWW	M	MW	MWW
	77.5%	21.3%	46.1%	39.3%	14.6%	26.7%	26.7%	46.7%
W-to-W ratio	0.17–0.2		0.12–0.19			0.07–0.14		
Roof angle	25°–35°		25°–35°			25°–35°		
Floors	3	2	2	3		2	Other	
	53.9%	40.4%	52.8%	37.1%		86.2%	13.8%	
Roof area	340 m^2		304 m^2			270 m^2		
Main function	Fruit and crop farming, viticulture, dairy farming		Dairy farming, cereals and potato			Dairy farming		
Layout	Viticulture function: in the same building/close to each other; dairy farming: at different altitude		In the same building/close to each other			In the same building/close to each other		
Diagram								
General view								
	Amplatz, Montagna [1]		Umer, Lasa [2]			Unterleiten, Valle Aurina [3]		

[1] Office of Architectural and Artistic Heritage, Autonomous Province of Bolzano - Alto Adige, http://www.provincia.bz.it/arte-cultura/beni-culturali/monumentbrowser-ricerca.asp?status=detail&id=16079; [2] Office of Architectural and Artistic Heritage, Autonomous Province of Bolzano - Alto Adige, http://www.provincia.bz.it/arte-cultura/beni-culturali/monumentbrowser-ricerca.asp?status=detail&id=15539; [3] Office of Architectural and Artistic Heritage, Autonomous Province of Bolzano - Alto Adige, http://www.provincia.bz.it/arte-cultura/beni-culturali/monumentbrowser-ricerca.asp?status=detail&id=13640.

- Discussion with consideration of qualitative results

Extensive studies showed that the choice of construction materials depends much on their availability and on cultural reasons [50,51]. Cultural influence was widely discussed for nationalistic purposes to trace and validate the geographical borders of different cultural regions [45]. The stone structure is deemed as a typical characteristic of Latin and Rhaetian-Romanzo influence, while wood is of Alpine Germanic influence. Dating back to the early Roman period, the Mediterranean colonialists, whose diet was based on bread, wine, and oil, tended to settle in areas suitable for these crops (low-altitude areas). The Germanic people were more dependent on milk and its derivatives, and they could, therefore, settle at higher altitudes [45]. This corresponds to what Roberti et al. [51] observed, whereby a higher elevation denoted a larger proportion of wood in a farmhouse.

In addition to cultural reasons, the material preference in climate zones shows a correlation with climate, although the construction custom is not necessarily determined by climate. Different people divided the land into areas of different agriculture use according to climate conditions, and the function of the farmhouses followed the agriculture need. The choice of the material relates to the functional layout of the farmhouses.

The oldest type of building layout is mentioned in "Lex baiuvariorum", which is called Haufenhof with multiple buildings [52] (Figure 10). The buildings were limited by construction techniques; thus, most of them were small with one function: the dwelling, the barn, the stable, the granary, the bath, and the kitchen. With technical progress, larger buildings became possible. Paarhof and Einhof evolved from Haufenhof (Figure 11). Paarhof represents the most common type of farm in the Alpine region. In the survey of South Tyrol, about 65% of rural buildings can be described as Paarhöfe [53]. Paarhof can adapt well to every terrain, even to steep ground [54]. By the term of Paarhof, a farm layout is described where the dwelling building and the farm building stand independently. In most Paarhof, the farm buildings are constructed with wood, while the dwelling buildings are built with masonry. Einhof is a farm where dwelling function and farm function are located under one roof. It represents 15% of total rural residences in South Tyrol [53]. Like the Paarhof, the farm space is generally constructed with wood. Dwelling spaces, especially the kitchen and living room, are constructed with masonry to prevent fire accidents.

Figure 10. Haufenhof in the Alpine village of Fane-Vals. (Whgler, "Fane-Alm Gesamtansicht", https://commons.wikimedia.org/wiki/File:Fane-Alm_Gesamtansicht.jpg, 2015)

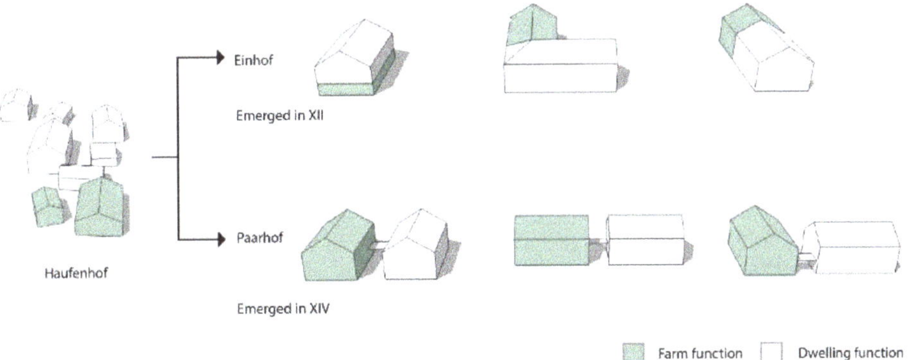

Figure 11. Farm forms and development.

The function of the farm greatly depends on the climate zone where it is located. In climate Zone I, grape and other fruits (especially apple) have a long tradition. According to the report of the BLS (Business Location Südtirol) [55] (Figure 12), they grow in areas from 200 to over 1000 m in altitude, extending westward from Val d'Adige to central Val Venosta (Malles Venosta), eastward until Valle Isarco (Natz). In the past, almost every farm worked independently in viticulture. Therefore, each farmstead possessed all the facilities required for wine production: a residential building, a stable, the torggel (room for winepress), and storage [56].

In climate Zone II, the main farm function is dairy farming. Currently, fruit planting dominates the western part of climate Zone II (Figure 12). However, it is only in the last 30 years that the domain changed from dairy farming to fruit planting in Val Venosta [56]. The same change also happened in the eastern part of climate Zone II (Valle Isarco). In climate Zone III, dairy farming is predominant due to the harsh climate for other agriculture. On dairy farms in climate Zones II and III, the attics are used as drying rooms for hay and agricultural products. Therefore, the construction of the attics uses wood with unglazed openings to ensure enough air exchange.

Figure 12. Agriculture and primary production in South Tyrol (© BLS/www.farbfabrik.it) [55].

Nature and culture both lead to farm function differentiation. Fruit, viticulture, and crop farming require the warm climate of the valley or on the south-facing slopes of the mountains, while the high Alpine pastures are suitable for grazing and dairy farming. Although the climate type determines the optimum land use, the actual use of the land depends more on the farmer's responses to economic opportunities [57]. Notably, people living at different altitudes engaged in both valley cultivation and mountain grazing, but with a different focus. This combined cultivation has a long history. Dating back to the Bronze Age, the transhumance system was found in South Tyrol [58]. For the settlements at low altitude, the function of dairy farming was placed on the mountain far away from the settlements. The stables and mountain huts were temporarily used in summer as a collective property.

The size of the residences may be influenced by the economic condition of the region. Another theory for the different sizes of farmhouses is that the depth of the house is commensurate with the length of the trees trunks available in the area [45]. The decreased window-to-wall ratio from climate Zone I to Zone III could be a climate-responsive feature that helps to decrease the energy loss through windows.

5.2.2. Portici House

- Description of quantitative results

According to the sample survey, all Portici houses have a similar layout: arcades facing the street, with shops occupying the ground floor and apartments located on the upper floors (Table 9). In Zone I, the shop and apartments extend toward the back, with an inner courtyard. In Zone II, on the other hand, a small yard is located behind the shop, leading to stables for livestock, with access from the back for staff and animals. The construction material is masonry in both climate zones. The dimension of the residence is larger in Climate Zone I than in Zone II, with average areas of roof projection of 447.6 m^2

and 360.1 m^2, respectively. The window-to-wall ratio is 0.21–0.4 in Zone I compared to 0.15–0.35 in Zone II.

Table 9. Construction of "Portici" buildings in different climate zones.

Climate Zone	Zone I	Zone II
Material	Masonry	Masonry
W-to-W ratio	0.21–0.4	0.15–0.35
Main facade	Eaves side	Gable side
Floors	See Table 7	See Table 7
Roof area	447.6 m^2	360.1 m^2
Average Width × depth	8.7 × 51.5 m	9.5 × 26.2 m
Layout type	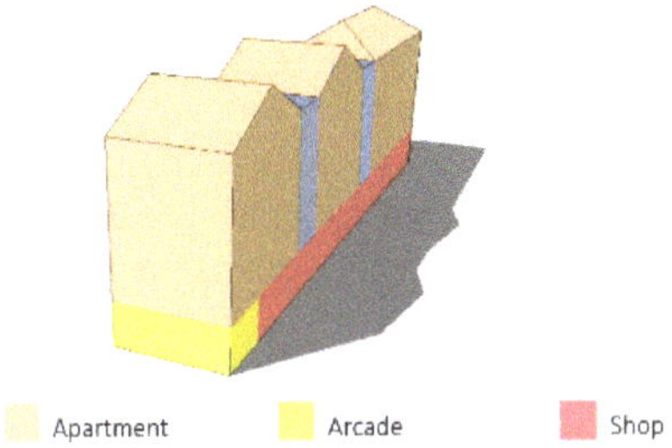	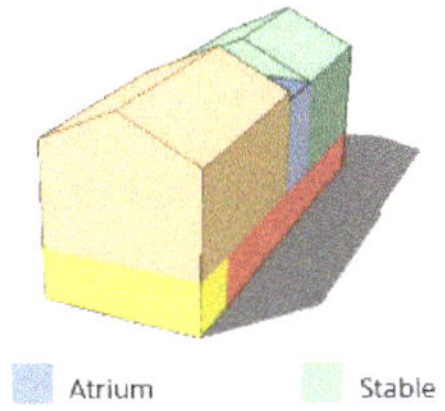
Typical diagram & Picture	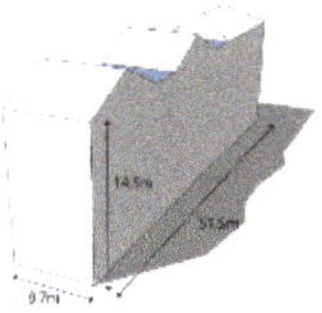  Portici 45, Bolzano [1]	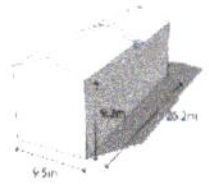  Via Portici 10, Glorenza [2]

[1] Office of Architectural and Artistic Heritage, Autonomous Province of Bolzano - Alto Adige, http://www.provincia.bz.it/arte-cultura/beni-culturali/monumentbrowser-ricerca.asp?status=detail&id=13862; [2] Office of Architectural and Artistic Heritage, Autonomous Province of Bolzano - Alto Adige, http://www.provincia.bz.it/arte-cultura/beni-culturali/monumentbrowser-ricerca.asp?status=detail&id=13862.

- Discussion with consideration of qualitative results

The differences in building layout and roof projection area are due to the development of trading activities. When they were initially constructed, Portici houses had a fixed layout in climate Zones I and II, with the shop facing the street and the stable at the back [48]. Portici houses developed due to the prosperity of the trading and craft. The stable was abandoned since the farm is no longer a main economic income, and the building extended backward. The depth of the extension could reach up to 60 m. To ensure enough light in the residence, two or three atria were inserted in between. Compared to the depth, the width of the building structure did not change much over time. In Bolzano, each parcel had a narrow, uniformed façade of about 6 m (about three windows wide), and 12 m for the duplex façade opening to the main street. This building structure had a very low surface-to-volume ratio (S/V) ratio. This compact structure ensured equal trading opportunities to as many shops as possible, saved public farmland and investments on original walls, and decreased the heat losses through the building envelope.

Building materials changed over time to increase fire safety. Every Portici district was seriously threatened by fire accidents. It is documented that the Portici houses in Bolzano were initially built in wood on upper floors [47]. Due to devastating fires, there was a large loss of property and lives. Masonry, therefore, became the preferred construction method for the following rebuild. In building samples, all the Portici buildings are in masonry.

6. Conclusions

The present paper proposes a methodology to categorize the historic building stock using a combination of quantitative and qualitative data. For the categorization, building features were analyzed on the settlement and building level. The analysis conducted allows highlighting building features that are influenced by climate and local culture, which contributes to the state of knowledge of the historic building stock. This methodology could be applied to different scales of historic building stock with the aim of understanding the correlation between building categories and climate.

The application of this methodology in South Tyrol shows the process in which a complex historic building stock is systematized. In addition to that, some correlations between building categories and local climate were discovered. From climate Zone I to III, the temperature decreases, and the precipitation increases as the altitude increases; the settlements of historic buildings tend to be sparser, with lower density; historic buildings tend to have smaller volumes, a lower window-to-wall ratio, less thermal mass, and different agriculture functions. These results not only show that settlements are more concentrated in regions with a climate that is ideal for agriculture, but also that they adapt to the climate in some ways. According to the analysis of the development of building features, climate is an important factor but not the only decisive one.

Considering future climate change, which could cause severe, pervasive, and irreversible impacts on historic buildings, it is necessary to study the performance of historic buildings to ensure their energy efficiency and conservation. According to the analysis of building features in three climate zones, it is necessary to use different archetypes to represent the typical buildings. Moreover, there is a need to carry out research to understand the capability of the climate-responsive features in future climate scenarios, as well as exploring the possible further risks and adaptation strategies.

Author Contributions: Conceptualization, A.T. and D.H.-A.; methodology, L.H. and D.H.-A.; data collection and analysis, L.H.; supervision, A.T. and C.D.P.; writing-original draft preparation, L.H.; writing-review and editing, D.H.-A., C.D.P. and A.T.

Funding: This research was funded by Thematic Scholarship "Energy Retrofit of Historic Buildings", grant number 32b-T4 and The APC was funded by Politecnico di Milano.

Conflicts of Interest: The authors declare no conflicts of interest.

References

1. Montalbán Pozas, B.; Neila González, F.J. Housing building typology definition in a historical area based on a case study: The Valley, Spain. *Cities* **2018**, *72*, 1–7. [CrossRef]
2. Famuyibo, A.A.; Duffy, A.; Strachan, P. Developing archetypes for domestic dwellings—An Irish case study. *Energy Build.* **2012**, *50*, 150–157. [CrossRef]
3. Hrabovszky-Horváth, S.; Pálvölgyi, T.; Csoknyai, T.; Talamon, A. Generalized residential building typology for urban climate change mitigation and adaptation strategies: The case of Hungary. *Energy Build.* **2013**, *62*, 475–485. [CrossRef]
4. *Alpine Building Culture and Energy Efficiency*; Alphouse: München, Germany, 2012.
5. IPCC. *Climate Change 2014 Synthesis Report*; IPCC: Geneva, Switzerland, 2015; p. 151.
6. EEA. *Climate Change, Impacts and Vulnerability in Europe 2016: An. Indicator-Based Report*; European Environment Agency: Gare, Luxembourg, 2017.
7. EURAC Research. *Klimareport Südtirol 2018*; Eurac Research: Bolzano, Italy, 2018.
8. Sesana, E.; Bertolin, C.; Gagnon, A.S.; Hughes, J.J. Mitigating Climate Change in the Cultural Built Heritage Sector. *Climate* **2019**, *7*, 90. [CrossRef]
9. Hao, L.; Herrera, D.; Troi, A. The effect of climate change on the future performance of retrofitted historic buildings: A review. In Proceedings of the EEHB, Visby, Sweden, 26–27 September 2018.

10. Vyazov, L.A.; Ershova, E.G.; Ponomarenko, E.V.; Gajewski, K.; Blinnikov, M.S.; Sitdikov, A.G. Demographic changes, trade routes, and the formation of anthropogenic landscapes in the middle Volga region in the past 2500 years. In *Socio-Environmental Dynamics Along the Historical Silk Road*; Springer: Berlin/Heidelberg, Germany, 2019; pp. 411–452.

11. Crate, S.A. Climate and culture: Anthropology in the era of contemporary climate change, in Annual Review of Anthropology. *Annu. Rev. Anthropol.* **2011**, *40*, 175–194. [CrossRef]

12. Hou, G.; Liu, F.; Xiao, L.; Zeng, Z.Z. The transmutation of settlements of prehistoric cultures in eastern Qinghai caused by climate change. *Acta Geogr. Sin.* **2008**, *63*, 34–40.

13. Rubio-Bellido, C.; Pulido-Arcas, J.; Cabeza-Lainez, J. Adaptation Strategies and Resilience to Climate Change of Historic Dwellings. *Sustainability* **2015**, *7*, 3695–3713. [CrossRef]

14. De Filippi, F. Traditional architecture in the Dakhleh Oasis, Egypt: Space, form and building systems. In Proceedings of the 23rd Conference on Passive and Low Energy Architecture (PLEA2006), Geneva, Switzerland, 6–8 September 2006.

15. Huang, L.; Hamza, N.; Lan, B.; Zahi, D. Climate-responsive design of traditional dwellings in the cold-arid regions of Tibet and a field investigation of indoor environments in winter. *Energy Build.* **2016**, *128*, 697–712. [CrossRef]

16. La Spina, V. *VerSus: Heritage for Tomorrow Vernacular Knowledge for Sustainable Architecture*; Firenze University Press: Firenze, Italy, 2014.

17. Alberti, F.; Chiapparini., C. (Eds.) *Cultura ed Ecologia Dell'architettura Alpina*; Regione del Veneto: Venice, Italy, 2012.

18. Bionaz, C. Analisi integrate per la conservazione e il miglioramento del comportamento energetico dei fabbricati rurali della Valle d'Aosta. *il Progetto Sostenibile-Ricerca e Tecnologie per l'ambiente Costruito* **2018**, *40*, 96–109.

19. Albatici, R. Local Tradition and Bioclimatic Architecture in the Italian Alpine Region. In Proceedings of the 23rd Conference on Passive and Low Energy Architecture (PLEA2006), Geneva, Switzerland, 6–8 September 2006.

20. *Decreto Legislativo 22 Gennaio 2004, n.42-Codice Dei Beni Culturali e del Paesaggio*; Gazzetta Ufficiale: Milano, Italy, 2004.

21. ISTAT. Data Warehouse of the 2011 Population and Housing Census. 2011. Available online: http://dati-censimentopopolazione.istat.it/Index.aspx?lang=en&SubSessionId=166061a9-96eb-4d4c-a6fc-3641b0124134&themetreeid=-200 (accessed on 24 November 2019).

22. EU. EU Building Database. Available online: https://ec.europa.eu/energy/en/eu-buildings-database (accessed on 24 November 2019).

23. Kavgic, M.; Mavrogianni, A.; Mumovic, D.; Summerfield, A.; Stevanovic, Z.; Djurovic-Petrovic, M. A review of bottom-up building stock models for energy consumption in the residential sector. *Build. Environ.* **2010**, *45*, 1683–1697. [CrossRef]

24. Filogamo, L.; Peri, G.; Rizzo, G.; Giaccone, A. On the classification of large residential buildings stocks by sample typologies for energy planning purposes. *Appl. Energy* **2014**, *135*, 825–835. [CrossRef]

25. Csoknyai, T.; Hrabovszky-Horváth, S.; Georgiev, Z.; Jovanovic-Popovic, M.; Stankovic, B.; Villatoro, O.; Szendrő, G. Building stock characteristics and energy performance of residential buildings in Eastern-European countries. *Energy Build.* **2016**, *132*, 39–52. [CrossRef]

26. Mahayuddin, S.A.; Zaharuddin, W.A.Z.W.; Harun, S.N.; Ismail, B. Assessment of Building Typology and Construction Method of Traditional Longhouse. *Proced. Eng.* **2017**, *180*, 1015–1023. [CrossRef]

27. Prieto, I.; Izkara, J.L.; Egusquiza, A. Building stock categorization for energy retrofitting of historic districts based on a 3d city model. *Dyna (Spain)* **2017**, *92*, 572–579. [CrossRef]

28. Genova, E.; Fatta, G.; Vinci, C. The Recurrent Characteristics of Historic Buildings as a Support. to Improve their Energy Performances: The Case Study of Palermo. *Energy Proced.* **2017**, *111*, 452–461. [CrossRef]

29. Santos, C.; Ferreira, T.M.; Vicente, R.; da Silva, J.R.M. Building typologies identification to support risk mitigation at the urban scale-Case study of the old city centre of Seixal, Portugal. *J. Cult. Herit.* **2013**, *14*, 449–463. [CrossRef]

30. Caniggia, G.; Maffei, G.L. *Architectural Composition and Building Typology: Interpreting Basic Building*; Alinea: Florence, Italy, 2001.

31. Ballarini, I.; Corgnati, S.P.; Corrado, V. Use of reference buildings to assess the energy saving potentials of the residential building stock: The experience of TABULA project. *Energy Policy* **2014**, *68*, 273–284. [CrossRef]
32. Dall'O', G.; Galante, A.; Torri, M. A methodology for the energy performance classification of residential building stock on an urban scale. *Energy Build.* **2011**, *48*, 211–219. [CrossRef]
33. Theodoridou, I.; Papadopoulos, A.M.; Hegger, M. A typological classification of the Greek residential building stock. *Energy Build.* **2011**, *43*, 2779–2787. [CrossRef]
34. Broström, T.; Bernardi, A.; Egusquiza, A.; Frick, J.; Kahn, M. A method for categorization of European historic districts and a multiscale data model for the assessment of energy interventions. In Proceedings of the 3rd European Workshop on Cultural Heritage Preservation (EWCHP), Bolzano, Italy, 16–18 Septemble 2013.
35. Broström, T.; Donarelli, A.; Berg, F. For the categorisation of historic buildings to determine energy saving. *Int. J. Archit. Art Des.* **2017**, *1*, 135–142.
36. ASTAT. *South. Tyrol in Figures-2018. Provincial Statistics Institute-Autonomous Province of South Tyrol*; ASTAT: Bolzano, Italy, 2018.
37. Adler, S.; Chimani, B.; Drechsel, S.; Haslinger, K.; Hiebl, J.; Meyer, V.; Resch, G.; Rudolph, J.; Vergeiner, J.; Zingerle, C.; et al. Das Klima: Von Tirol-Sudtirol-Belluno, ZAMG, Autonome Provinz Bozen, ARPAV, Eds. 2015. Available online: http://www.3pclim.eu/images/Das_Klima_von_Tirol-Suedtirol-Belluno.pdf (accessed on 24 November 2019).
38. Köppen, W. The thermal zones of the Earth according to the duration of hot, moderate and cold periods and to the impact of heat on the organic world. *Meteorol. Z.* **2011**, *20*, 351–360. [CrossRef] [PubMed]
39. ASTAT. *Statistisches Jahrbuch 2017/Annuario statistico 2017: 15 Bautätigkeit und Wohnungen-Edilizia e Abitazioni*; ASTAT: Bolzano, Italy, 2017.
40. *Hygrothermal Performance of Buildings—Calculation and Presentation of Climatic Data Part. 3: Calculation of a Driving Rain Index for Vertical Surfaces From Hourly Wind and Rain Data*; EN 15927-3; EU: Bruxelles, Belgium, 2009.
41. GeoCatalogo-Rete Civica dell' Alto Adige. 2018. Available online: http://geokatalog.buergernetz.bz.it (accessed on 24 November 2019).
42. Tinner, W.; Lotter, A.F.; Ammann, B.; Conedera, M.; Hubschmid, P.; van Leeuwen, J.F.; Wehrli, M. Climatic change and contemporaneous land-use phases north and south of the Alps 2300 BC to 800 AD. *Quat. Sci. Rev.* **2003**, *22*, 1447–1460. [CrossRef]
43. Büntgen, U.; Bellwald, I.; Kalbermatten, H.; Schmidhalter, M.; Frank, D.C.; Freund, H.; Bellwald, W.; Neuwirth, B.; Nüsser, M.; Esper, J. 700 years of settlement and building history in the Lötschental, Switzerland. *Erdkunde* **2006**, *60*, 96–112. [CrossRef]
44. Pessina, C.; Tronconi, O. *L'Architettura Montana*; Maggioli Editore: Santarcangelo di Romagna, Italy, 2008.
45. Curzel, V. *Living the Alps Ecological Architecture, Development Models, Identity Buildings: The cases of Alto Adige Südtirol and Trentino, in Dipartimento di Filosofia, Sociologia, Pedagogia e Psicologia Applicata-Fisppa*; Universita degli studi di Padova: Padua, Italy, 2013.
46. Büntgen, U.; Esper, J.; Frank, D.C.; Nicolussi, K.; Schmidhalter, M. A 1052-year tree-ring proxy for Alpine summer temperatures. *Clim. Dyn.* **2005**, *25*, 141–153. [CrossRef]
47. Trentini, C. *Von Pons Drusi zu Bozen-Ikonographie und Iknographie der Stadt Bozen*; Edition Sturzflüge: Bolzano, Italy, 1996.
48. Exner, D.; Lucchi, E. Learning from the past: The recovery and the optimization of the original energy behaviour of "Portici" Houses in Bolzano. In Proceedings of the 49th International Conference, Historical and Existing Buildings: Design the Retrofit (AICARR), Rome, Italy, 26–28 February 2014.
49. Pertoll, A.P. *Ins Licht Gebaut-Die Meraner Villen*; Edition Raetia: Roma, Italy, 2009.
50. Gerardo-Unia, G.D. *Abitare le Alpi*; L'Arciere: Cuneo, Italy, 1980.
51. Roberti, F.; Exner, D.; Troi, A. Energy Consumption and Indoor Comfort in Historic Refurbished and Non-refurbished Buildings in South Tyrol: An Open Database. In Proceedings of the Smart and Sustainable Planning for Cities and Regions, Bolzano, Italy, 22–24 March 2017.
52. Konrad, B. *Natürliche Bauweisen: Bauernhöfe in Südtirol*; Athesia Spectrum: Bolzano, Italy, 2008.
53. Jones, G.A.; Warner, K.J.; France, N. The 21st century population-energy-climate nexus. *Energy Policy* **2016**, *93*, 206–212. [CrossRef]
54. Pohler, A. *Die schönsten Bauernhöfe in Tirol: Nordtirol-Osttirol-Südtirol*; Tyrolia Verlagsanstalt Gm: Innsbruck, Austria, 2007.

55. BLS. *Südtirol-Lebensmitteltechnologien*; BLS: Bolzano, Italy, 2015.

56. Laimer, M.; Dellagiacoma, R. Bauen Im Ländlichen Raum-Beispiele Bestehender Hof-und Architekturtypologien in Südtirol. Available online: http://www.provinz.bz.it/natur-umwelt/natur-raum/publikationen.asp?publ_action=300&publ_image_id=9701 (accessed on 24 November 2019).

57. Lambin, E.F.; Turner, B.L.; Geist, H.J.; Agbola, S.B.; Angelsen, A.; Bruce, J.W.; Coomes, O.T.; Dirzo, R.; Fischer, G.; Folke, C.; et al. The causes of land-use and land-cover change: Moving beyond the myths. *Glob. Environ. Chang.* **2001**, *11*, 261–269. [CrossRef]

58. Putzer, A.; Festi, D.; Edlmair, S.; Oeggl, K. The development of human activity in the high altitudes of the Schnals Valley (South. Tyrol/Italy) from the Mesolithic to modern periods. *J. Archaeol. Sci. Rep.* **2016**, *6*, 136–147. [CrossRef]

Article

Climate Change and Sustaining Heritage Resources: A Framework for Boosting Cultural and Natural Heritage Conservation in Central Italy

Ahmadreza Shirvani Dastgerdi [1,*], Massimo Sargolini [1], Shorna Broussard Allred [2], Allison Chatrchyan [3] and Giuseppe De Luca [4]

[1]   School of Architecture and Design, University of Camerino, 63100 Ascoli Piceno, Italy; massimo.sargolini@unicam.it

[2]   Center for Conservation Social Sciences, Department of Natural Resources & Southeast Asia Program, Cornell University, Ithaca, NY 14853, USA; srb237@cornell.edu

[3]   Department of Development Sociology, Cornell University, Ithaca, NY 14853, USA; amc256@cornell.edu

[4]   School of Urban and Regional Planning, University of Florence, 50121 Firenze, Italy; giuseppe.deluca@unifi.it

*   Correspondence: ahmadreza.shirvani@unicam.it

Received: 9 January 2020; Accepted: 31 January 2020; Published: 5 February 2020

Abstract: Climate change has dramatically affected the rainfall patterns and water systems in Central Italy. The vulnerability of this area to climate change and natural hazards necessitates that appropriate adaptation policies be put in place to protect heritage sites. This study aims to develop a cultural and natural heritage conservation framework for Central Italy that enhances the capacity of climate change adaptation for heritage resources. For this purpose, a comparison was made between the UNESCO (United National Educational, Scientific and Cultural Organization) Convention of 1972 and the European Landscape Convention of the Council of Europe to achieve a coherent vision for the protection of heritage resources in Europe. After describing the impacts of climate change on heritage resources in Central Italy, we analyze and suggest improvements to the conservation framework for wisely protecting heritage resources in a changing climate. The findings reveal that conservation sectors require assessments of the value of heritage resources at the territorial scale to effectively define conservation priorities, assess the vulnerabilities, and more precisely direct funding. In this respect, the integration of the European Landscape Convention with territorial planning may boost the unity of a conservation framework in terms of climate change while providing new opportunities for conservation authorities to develop adaptation policies.

Keywords: climate change; heritage resources; conservation; adaptation capacity; territorial planning

1. Introduction

Heritage resources include all cultural and natural sites, and in particular, World Heritage Sites. As designated by UNESCO (United National Educational, Scientific and Cultural Organization), World Heritage Sites are a type of "cultural landscape" that conserve cultural and natural heritage sites around the world [1]. Each World Heritage site is designated on the basis of one or more Outstanding Universal Values [2], which are assessed through a rigorous evaluation process by the Advisory Bodies of the World Heritage Convention [3]. According to article 49 of the convention, Outstanding Universal Value is defined as "cultural and/or natural significance which is so exceptional as to transcend national boundaries and to be of common importance for present and future generations of all humanity" [4]. As a type of cultural landscape, World Heritage Sites prominently boost the economy of countries by attracting numerous domestic or international visitors [5]. From a broader perspective, Plachter and Rossler (1995) define cultural landscapes as the result of "the interactions between people and their natural environment over space and time" [1]. In cultural landscapes, heritage resources recall historical identities and enhance social cohesion [6].

From a territorial planning point of view, there is a strong tie between the proper conservation of heritage resources and the sustainability of cultural landscapes [7–9]. Heritage resources were often created based on a stable climatic condition [10]. However, global climate change—which is causing increases in extreme weather events—has emerged as a significant threat to the sustainability of many heritage resources all over the world [11,12].

Climate change inflicts direct damage on the materials and structure of historical monuments [13] and threatens the status of Outstanding Universal Value at many World Heritage Sites [14]. For instance, the United Kingdom Climate Impacts Programme (UKCIP) has suggested that the sea level rise in the Thames Estuary will reach between 0.26 and 0.86m by the year 2080, compared to its average level between 1961 and 1990 [15]. Such a rise in sea level and projected changes in storm patterns will pose a significant threat to the Outstanding Universal Value of three World Heritage sites in London, including the Tower of London, the Palace of Westminster, and Maritime Greenwich [10]. Furthermore, climate change has indirect impacts, such as severe weather events that diminish the number of visitors [16] and disrupt socio-economic activities at cultural landscape areas [17]. For example, climate change is projected to decrease the annual visitations numbers of the Mesa Verde National Park in the USA, which attracts about 500,000 tourists yearly and contributes about US$ 47 million to the local economy [18].

According to Intergovernmental Panel on Climate Change (2014), adaptation is the process of adjustment to actual or expected climate and its effects. In human systems, adaptation seeks to moderate or avoid harm while harnessing beneficial opportunities [19]. Thus, climate adaptation involves the actions taken to address the impacts of climate change by reducing vulnerability and taking advantage of opportunities. Climate adaptation actions have the potential to reduce governmental barriers and create an environment in which private sector actors can invest in adaptation strategies fit for a future without carbon [20]. Many cities and mega-cities have improved their climate adaptation capacity through policymaking and have developed adaptation measures for reducing vulnerability to the effects of climate change [21]. In this respect, Dilling (2017) explains how cities in the USA have successfully used policymaking to cope with climate and water variability and increasing population [22].

Building adaptation capacity may be defined at various levels (i.e., international, national, or local), and subnational jurisdictions are essential sectors for improving measures to reduce the risks of climate change [23]. The adaptation capacity at each level plays a vital role in the effectiveness of adaptation policies in the long-term [24].

Heritage resources are easily vulnerable to climate-related risks, such as heavy rainfall events, or landslides (UNESCO, 2008). A proper understanding of climate impacts on heritage resources not only provides a reliable base for formulating and developing adaptation policies, but it can also help to reduce the costs of climate change impacts, by building adaptive capacity to conserve those resources (Galeotti and Roson, 2012). Therefore, an inter-disciplinary framework is needed to effectively addresses the complex interactions between climate, physical, social, and ecological systems [25]. Delayed action in the present may diminish opportunities for adaptation pathways in the future [26]. Many cultural heritage experts and researchers in Europe believe that the adaptation of heritage resources to climate change is possible [27].

Local case studies that analyzed the adaptive capacity of a particular region or a community argue for the need to assess and measure adaptive capacity at the regional or local level because the decisions to adapt are often made at that level [28,29]. Building on the IPCC's definition of adaptive capacity, Smit and Pilifosova classify the determinants of adaptive capacity in six categories, consisting of Economic Resources, Technology, Infrastructure, Information and Skills, Institutions, and Equity [30]. It is widely accepted that economic assets, capital resources, financial means and wealth all play an important role in adaptive capacity. Wealthy nations are more likely to be in a better position to adapt to changes in the climate and can bring more resources to bear on the costs of adaptation. Technological resources enable adaptation options, and consequently the lack of access and development of technology can lead to lower adaptive capacity. Existence and development of infrastructure can form the basis for the development of adaptation options and measures. Skilled, informed, and trained personnel enhance adaptive capacity and access to information is likely to lead to development of adaptation options that are timely and appropriate. Well-developed institutions and governance structures not only have the capacity to address present day challenges but also provide a decision-making infrastructure that enables communities to effectively plan for future. Finally, equity is a measurement tool for social sensitivity to climate change and pays needed attention to populations most vulnerable to climate change impacts [31]. Some studies also highlight the necessity of developing communication mechanisms among heritage institutions, academic researchers, and the local community to build adaptation capacity at heritage sites [13,32]. Communication is essential for adapting to climate change because it facilitates the engagement of the public in climate change science and solutions in partnership with governments, media organizations, companies, and civil society [33].

In Italy, heritage resources play a significant role in social cohesion as markers of historical identity and are also a driving factor in the national economy. In this respect, The Italian National Institute of Statistics affirms that in 2017 visitors spent about 421 million nights at tourist accommodations, which represents an increase of 4.4% in comparison with the past year. However, few studies warn about the adverse economic impacts of climate change on Central Italy's heritage tourism. Yet the climate is changing; for example, climatic data shows that 2008 was recorded as the third wettest year in Italy since 1961 [34]. Thus, urgent attention is needed to develop mitigation and adaptation policies that take heritage resources into account.

Tourism and agricultural production are the most vulnerable economic sectors in Central Italy to climate change [35]. A recent World Bank report also highlights Italy's considerable vulnerability to climate change in the coming decades [36]. Heritage resource conservation remains a concern due to extreme events (droughts, flooding, and landslides), which are increasing in Central Italy as they are in many other regions of the world [37]. Therefore, the long-term sustainability of heritage resources in Central Italy requires a robust conservation framework that enhances the adaptation capacity in this region. This study aims to develop a resilient framework that can be used to assess the adverse climate impacts on cultural and natural heritage sites in Central Italy, and bring new opportunities for adaptation of heritage resources to climate change.

2. Research Design

2.1. Materials and Methods

This study adopts a descriptive qualitative comparative approach for boosting the current cultural and natural heritage conservation framework in Central Italy. For this purpose, we first compare two heritage conservation frameworks in Italy—the UNESCO Convention of 1972 and the European Landscape Convention of the Council of Europe—to achieve a coherent vision for the protection of heritage resources in Europe, and in particular, Italy. In the results, the impacts of climate change on various heritage resources are discussed to provide an understanding of the conservation challenges in Central Italy. Finally, we propose a roadmap for promoting the cultural and natural heritage conservation framework and building capacity for adaptation of heritage resources to climate change in this area.

2.2. Study Area

Italy is characterized by a very complex climatic formation due to the presence of high mountain ranges, such as the Alps and Apennines, and the vicinity of the Mediterranean Sea. Regarding the atmospheric circulation, the Mediterranean Sea and land area of Italy are generally influenced by tropical air masses in the summer and by western air masses in winter. The variability of these circulation patterns and the interactions with such a complex system make the Mediterranean region especially vulnerable to climate change and sensitive both to the global phenomena and to events at the local scale [38].

This study focuses on Central Italy, which is one of the five macro regions of the country, with a total area of 58,085 km^2. Central Italy is composed of the four regions of Tuscany, Lazio, Umbria, and Marche [39]. Central Italy is dominated by the hills and mountains of the Apennines, from which a few major rivers flow. There are few natural plains of any size in this region, but those that do exist are famously fertile. They have been supplemented over the years by a process of land reclamation that has turned the coastal swamps and marshes into highly productive agricultural land, and provided space for the expansion of cities and towns [40]. With 55 World Heritage Sites, Italy is a leader under the UNESCO World Heritage List, and Central Italy hosts 13 World Heritage Sites (Figure 1 and Table 1) [41].

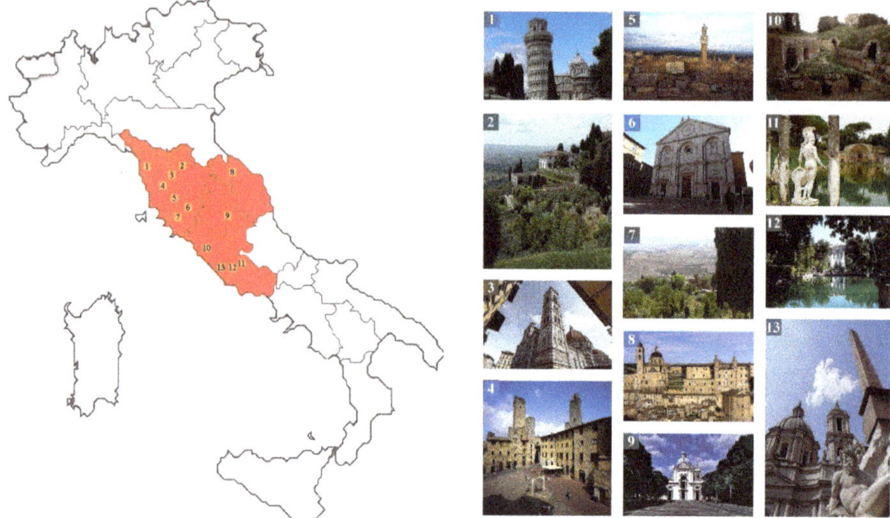

Figure 1. World heritage sites in Central Italy [42].

Table 1. World Heritage Sites in Central Italy [43].

Pr.	World Heritage Site	Year of Inscription	Category of Inscription
	1. Piazza del Duomo, Pisa	1987	(i)(ii)(iv)(vi)
	2. Medici Villas and Gardens	2013	(ii)(iv)(vi)
Tuscany	3. Historic Centre of Florence	1982	(i)(ii)(iii)(iv)(vi)
	4. Historic Centre of San Gimignano	1990	(i)(iii)(iv)
	5. Historic Centre of Siena	1995	(i)(ii)(iv)
	6. Historic Centre of Pienza	1996	(i)(ii)(iv)
	7. Val d'Orcia	2004	(iv)(vi)
Marche	8. Historic Centre of Urbino	1998	(ii), (iv)
Umbria	9. Assisi, the Basilica of San Francesco and Other Franciscan Sites	2000	(i), (ii), (iii), (iv), (vi)
	10. Etruscan Necropolises of Cerveteri and Tarquinia	2004	(i)(iii)(iv)
Lazio	11. Villa d'Este	2001	(i)(ii)(iii)(iv)(vi)
	12. Villa Adriana	1999	(i)(ii)(iii)
	13. Historic Centre of Rome	1980	(i)(ii)(iii)(iv)(vi)

Many of these sites are valuable treasures of the medieval and Renaissance era (Table 1), which meet different criteria of Outstanding Universal Value (Table 2).

Table 2. The Criteria for Outstanding Universal Value Based on the Operational Guidelines for the Implementation of the World Heritage Convention [4].

Type	Criteria	Definition
Cultural	(i)	to represent a masterpiece of human creative genius
	(ii)	to exhibit an important interchange of human values, over a span of time or within a cultural area of the world, on developments in architecture or technology, monumental arts, town-planning or landscape design
	(iii)	to bear a unique or at least exceptional testimony to a cultural tradition or to a civilization which is living or which has disappeared
	(iv)	to be an outstanding example of a type of building, architectural or technological ensemble or landscape which illustrates (a) significant stage(s) in human history
	(v)	to be an outstanding example of a traditional human settlement, land-use, or sea-use which is representative of a culture (or cultures), or human interaction with the environment especially when it has become vulnerable under the impact of irreversible change
Natural	(vi)	To be directly or tangibly associated with events or living traditions, with ideas, or with beliefs, with artistic and literary works of outstanding universal significance.
	(vii)	to contain superlative natural phenomena or areas of exceptional natural beauty and aesthetic importance
	(viii)	to be outstanding examples representing major stages of earth's history, including the record of life, significant on-going geological processes in the development of landforms, or significant geomorphic or physiographic features
	(ix)	to be outstanding examples representing significant on-going ecological and biological processes in the evolution and development of terrestrial, fresh water, coastal and marine ecosystems and communities of plants and animals
	(x)	to contain the most important and significant natural habitats for in situ conservation of biological diversity, including those containing threatened species of Outstanding Universal Value from the point of view of science or conservation

3. Analysis and Results

3.1. Landscape as a Cultural Heritage

The strong tie between cultural and natural heritage sites is evident in many "cultural landscapes" in Italy, and in particular, in Central Italy where there is a harmonious combination between many small historic centers and their natural heritage and natural resources. For instance, the landscape of Val d'Orcia in the province of Siena in Tuscany is part of the agricultural hinterland of Siena, redrawn and developed when it was integrated into the territory of the city-state in the 14th and 15th centuries to reflect an idealized model of good governance. The landscape of the Val d'Orcia was celebrated by painters from the Siennese School, which flourished during the Renaissance. Images of the Val d'Orcia, and particularly depictions of landscapes where people are drawn as living in harmony with nature, have come to be seen as icons of the Renaissance and have profoundly influenced the development of landscape thinking [44].

The palace and the gardens of Villa d'Este in Tivoli, in the center of Italy, is another example of this harmonious combination. The Villa d'Este had a profound influence on the development of garden design throughout Europe. The gardens with the fountains are a masterpiece of hydraulic engineering, both for the general layout of the plan and the complex system of distribution of water. A broad area around the Villa is protected as landscape by a law decree (Decreto n.42/04) to be of interest for the landscape resources [45].

To promote the meaning of cultural heritage, the Italian Ministry of Cultural Heritage and Activities adopts a "landscape vision" as a mixture of cultural goods and landscape resources [46]. As climate change impacts the features of the landscape, the cultural and natural heritage conservation framework should be able to assess the importance and vulnerability of heritage resources at the early stages of communication and policy making [25]. Despite having a mutual goal of heritage conservation, the approaches of the two conventions (UNESCO Convention of 1972 and Europe Landscape Convention) are distinct, which leads to different definitions of cultural and natural landscapes, and consequently, different heritage policies.

3.2. Approaches for Cultural Landscape

In the UNESCO 1972 Convention, the heritage concept is divided into two independent and separate notions of cultural and natural heritage. According to Articles one and two of the UNESCO Convention, cultural heritage is defined as monuments and groups of buildings and sites that have historical, aesthetic, archaeological, scientific, ethnological, or anthropological value. Natural heritage is defined as exceptional physical, biological, and geological formations, habitats of threatened animal and plant species, and areas that have scientific or aesthetic value [47]. The cultural criteria were slightly revised to the "cultural landscape" in the Operational Guidelines for the Implementation of the World Heritage Convention in 1992, albeit the origin of the concept harkens back to Carl O. Sauer in 1925 [48]. As shown in Table 3, UNESCO categorizes cultural landscapes into three types [49]. In the vision of UNESCO, historical features are a vital factor for the maintenance and inscription of a cultural landscape on the World Heritage list. For instance, the presence of traditional crops and local products, as well as the presence of architecture related to agricultural activities, are considered by UNESCO to be the significant factors of integrity in a cultural landscape. However, problems arise in determining how best to quantify the threshold of traditional features or historical durability in a changing climate. Furthermore, maintaining the integrity of the heritage sites is strongly influenced by social and economic factors.

Table 3. UNESCO Categories for Cultural Landscape.

Cat.	Type of Cultural Landscape	Definition
(i)	Clearly defined	Designed and created intentionally by man. This embraces garden and parkland landscapes characteristically constructed for aesthetic, social and recreational reasons which are often associated with religious or other monumental buildings and ensembles
(ii)	Organically evolved	Fall into two sub-categories. First, a relict landscape is one in which an evolutionary process came to an end at some time in the past. Second, is one which retains an active social role in contemporary society closely associated with a traditional way of life.
(iii)	Associative cultural	Landscapes with definable powerful, religious, artistic, or cultural associations with the natural element rather than material cultural evidence.

The European Landscape Convention—which was adopted in Florence in 2000—is the first international agreement exclusively dedicated to the cultural landscape. It describes the cultural landscape as any part of the territory "as perceived by people, whose character is the result of the action and interaction of natural and human factors" [50]. The convention emphasizes that landscape is derived from the interaction of a community and its environment over time. The vision of the European Landscape Convention is different from the UNESCO Convention, because its approach to cultural landscape is a quality of each territory, not only of those with exceptional landscape values. While the European Landscape Convention is more regionally focused and takes into consideration a holistic and social landscape, the UNESCO

Convention seeks to protect only those landscapes which are recognized worldwide as an exceptional heritage [51].

Before developing disconnected climate adaptation policies, there is value in determining what distinguishes buildings, landscapes, and other aspects of community into a unique culture. Furthermore, the cultural landscape is not so much something that is particularly beautiful or exceptional, but rather different and unique. Thus, it is necessary to sensitize and sufficiently educate both governments and communities about the value of cultural landscapes and involve them in heritage planning [52]. This approach gathers assets of agricultural and forestry interests, architectural assets, and cultural traditions in a distinctive or characteristic set. The cultural landscape approach meets both natural heritage and cultural heritage, within which we must consider not only the agricultural or forest heritage, but also the architectural aspects, including archaeological sites [53]. Therefore, adopting the approach of the European Landscape Convention is more fitting for defining a more grounded concept of cultural heritage in Central Italy and thus building climate adaptation capacity in this area [54].

3.3. Climate Change Impacts in Central Italy

Climate change and its related natural risks can disrupt the socio-economic dynamics in cultural landscapes. According to the Italian National Institute of Statistics [55], the average annual temperature of the country over the time period from 2002–2016 was 15.5 °C, an increase of 1.0 °C when compared to the years 1971-2000. In the period from 2002-2016, there were, on average, 110 summer days and 45 tropical nights; this is, respectively, 17 and 14 more than the climatological average for Italy. Climate observations confirm an increase in the average monthly temperature as well as an upward trend in extreme temperatures (Figure 2). In Tuscany, for instance, the maximum temperature increase (+0.44 °C/decade) was slightly higher than minimum temperature (+0.38 °C/decade) and, consequently, an increase in summer daily temperature range was noted (+0.06 °C/decade) [56].

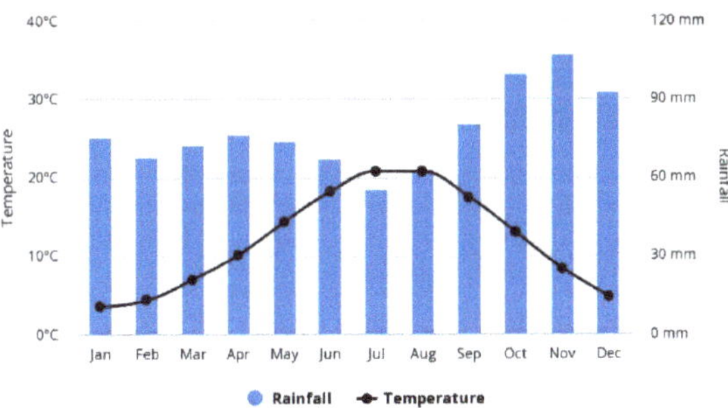

Figure 2. Average monthly temperature and precipitation in Italy from 1901-2016 [57].

In Central Italy, floods and landslides are among the leading risks that may impose irreparable damage to cultural landscapes (Figure 3). In this sense, the intangible characteristics of the cultural landscape may be threatened by loss of inhabitants due to damage to infrastructure, the spread of pollution and water-related diseases, soil erosion, and damage to agricultural lands and crops. For instance, between 1939 and 2004, Italy was hit by 28 massive floods that affected 2.85 million people and caused 1.5 million to become homeless. The number of victims over this time period was 694 people, and the cost of these climate impacts was estimated at US$ 32.7 million. After the floods of Campania in May 1998, climate risks were exacerbated by naturally occurring disasters in the region, such as the 2016 earthquake in Central Italy that devastated heritage resources and claimed the lives of nearly 300 people. In response, Italy developed a national plan "Urgent measures for the prevention of hydrogeological risk and for areas affected by landslides in the Campania region" under the law 267/98 for reducing the risks of floods and landslides [58]. However, this law does not take into account the higher risks deriving from climate change scenarios, for which no assessment currently exists [59].

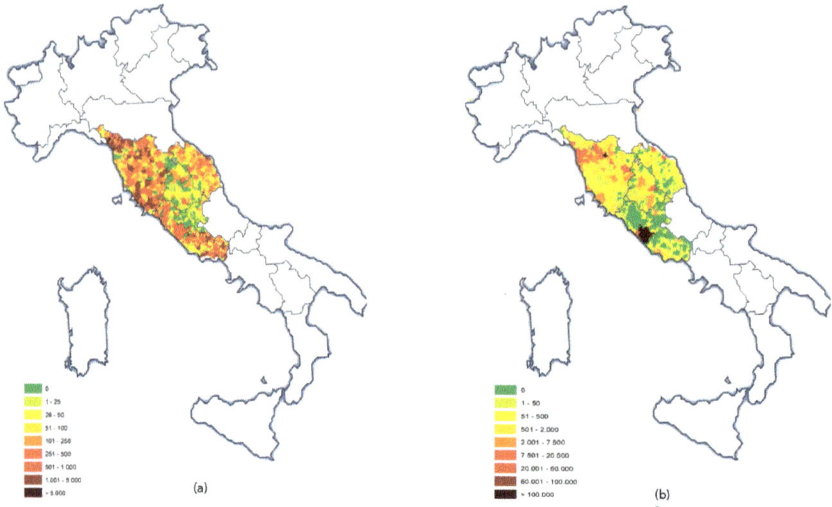

Figure 3. (**a**) The population at risk of landslides in Central Italy; (**b**) The population at risk of floods in Central Italy, derived from [60].

The risks of severe weather patterns caused by climate change may lead to irreversible damages to the value of many heritage resources in Central Italy. These threats affect not only the identity of the landscape but also the tourism industry. The tourism activity and economy may be interrupted or diverted, at least until livable conditions are restored in areas affected by floods or landslides.

Another concern is the loss of biodiversity and natural ecosystems due to the impacts of climate change. Natural resources are an integral part of the identity of territories in Italy. Moreover, in Central Italy, cultural landscapes encompass many of Natura 2000 sites, which include the most valuable and threatened species and habitats in Europe. In the region of Umbria, for instance, Natura 2000 sites cover about 15% of the entire territory. The sites encompass 41 habitats of European interest, of which 11 are defined as rare species, with 143 animal species (four are defined as primary), and eight species of plants [61].

The Italian public is mostly aware of the seriousness of climate change, but climate change is considered less urgent than other matters of concern related to the economic situation [62], specifically in terms of conservation policies. In developing adaptation capacity in cultural landscapes, Italian citizens rely mainly on information provided by traditional mass media, while environmental organizations and public communication by scientists play a marginal role. Some studies developed cyberinfrastructure models for disaster management [63].

4. Discussion

The current vision in Central Italy for conservation and management of heritage resources is still very close to the traditional paradigm that affords tangible heritage areas an intrinsic value, worthy of great efforts for its collection, catalogue and preservation in view of the likely restoration of the damaged heritage [64]. However, both protection of heritage areas and climate change adaptation are dynamic fields of investigation [65]. Prior to developing climate adaptation policies, it is necessary to examine the meaning of cultural heritage in light of climate change. According to the results, viewing heritage resources with a broader lens of the cultural landscape provides a comprehensive framework for the wise development of adaptation policies. In the vision of UNESCO, historical features play a vital role in the definition of the cultural landscape and can be divided into three different categories, including (1) clearly defined; (2) organically evolved; and (3) associatively cultural.

In contrast, the European Landscape Convention focuses on the social values of the cultural landscape that make a community distinct and unique. In this vision, the cultural landscape provides a holistic perspective—an area, as perceived by people, whose character is the result of the action and interaction of human factors and natural resources therein. Furthermore, in this vision, the local economy is linked to landscape resources in several ways: through resources related to the production of food, energy, raw material, and water (farming, forestry, fishery, water supply) and through the tourism industry. The European Landscape Convention promotes a valuation system based on community perception. It also pays particular attention to the strategic role of cultural and natural goods in sustaining the cultural landscape. From a territorial planning point of view, cultural heritage may be defined as a system of synergistic relationships between unique qualities of the physical environment, the built environment, and the anthropic environment [66]. Therefore, adaptation policies should be developed under a cultural and natural heritage conservation framework that suitably recognizes these three aspects in an integrated form.

The analysis of climate change impacts on Central Italy (Figure 3) illustrates the varying vulnerability to natural risks across the landscape. From a territorial planning point of view, developing climate-adaptive policy and reducing the impacts of natural risks is not only about conservation of the physical features in a heritage site, but also recognizing and protecting socio-economic activities that are linked to the cultural, natural, or World Heritage Sites. The socio-economic changes associated with climate impacts can have a significant influence on the protection of the cultural landscape. For identification of climate change impacts on socio-economic dynamics, heritage sectors need to estimate the damage costs of climate change on the cultural landscape and be fully aware of how this damage can be minimized by adaptation policies that build capacity and resilience. Such a vision would enhance the concept of cultural heritage from an isolated site into a territorial resource.

To promote the adaptation capacity of cultural landscapes, it is recommended that Central Italy integrate the European Landscape Convention within territorial planning systems. Such a combination provides an opportunity to use multidisciplinary and multi-sector perspectives in dealing with protection challenges, while also considering the unique morphologies of cultural landscapes in this area. For instance, if agriculture factors centrally in the identity of a cultural landscape, understanding farmers' views, framing issues, and necessary actions would be crucial for developing climate adaptation policies [54]. Adopting a territorial vision in a cultural and natural heritage conservation framework facilitates the inclusion of new approaches for adaptation of heritage resources in the face of climate change. The proposed approach supports multi-scale and multi-sector actions rooted in the differing expectations of a wide range of partners and expands the scope of community participation and engagement.

Consequently, territorial planning facilitates communication between planning institutions, stakeholders, and researchers in developing adaptation policies and reducing the risks of climate change. Furthermore, in this vision, cultural and natural heritage goods are dynamically assessed as useful characteristics for the sustainability of the cultural landscape. Adopting a territorial approach to heritage resources from conservation to protection, static to dynamic, valued to valuing, and from isolated to contributor can play a fundamental role in adaptation to climate change in Central Italy.

The proposed cultural and natural heritage conservation framework (Figure 4) not only supports mitigation and adaptation strategies at the national scale but also boosts communication among heritage institutions, academic researchers, and the local community. However, developing and implementing adaptation policies in such a framework highlights the need for further research to address the potential challenges and risks. Although this model affirms that losing a part of heritage resources because of climate change impacts is inevitable, inadequate knowledge of how vulnerabilities and the cost of climate change should be prioritized in the context of a cultural landscape remains a challenge for developing climate-adaptive policies.

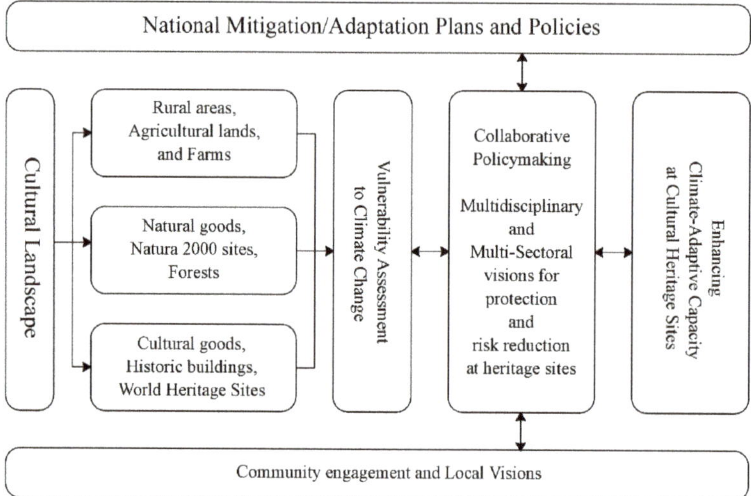

Figure 4. A framework for building adaptation capacity and sustaining of cultural landscapes in Central Italy.

Another consideration is that conservation organizations are at a different level of preparedness in terms of how well they are equipped to respond to climate change, and this condition can considerably affect the assessment, policymaking, and implementation steps. For instance, while local governments are the first responders for climate-related natural risks (e.g., flood), many municipalities may not be adequately prepared for dealing with floods. Most municipal officials have a reasonable understanding of flood issues and its adverse impact on cultural landscapes. However, they require additional information, support, and education as they deal with the complexity of decisions surrounding climate change and its impacts on heritage resources [67].

Considering the complexity of conservation issues in Central Italy in terms of climate change, the cultural and natural heritage conservation framework requires a model of how the exchanging of information should be developed among citizens, public administrators, and the scientific community at the regional scale.

5. Conclusions

The analysis of historical climate data shows that climate change has notably affected temperature and rainfall patterns in Italy. Many international reports forecast that Italy is going to be more vulnerable to climate change over the coming decades. In Italy, heritage resources are a driving factor for social cohesion and the national economy; thus maintenance of these resources necessitates urgent attention in developing adaptation policies at heritage sites. This study examined methods for strengthening the adaptation capacity at cultural and natural heritage sites in Central Italy. Our analysis revealed that floods and landslides are two significant climate-related risks that threaten Central Italy. However, different municipalities have different vulnerability levels to these risks based on their information and skills, and their institutional preparedness. This study further develops the concept of cultural heritage as a system of synergistic relationships between the unique qualities of the physical environment, the built environment, and the anthropic environment. Therefore, climate adaptation policies must integrate these three aspects to be effective at conserving and protecting cultural and natural heritage. In this respect, we propose the European Landscape Convention as a potential conservation framework for the development of adaptation policies in Central Italy. This convention takes a regional approach to cultural heritage that makes it distinct from the UNESCO vision. The European Landscape Convention takes the holistic and social landscape into consideration, while the approach of UNESCO is less regional and less place-specific in its focus. Integrating the European Landscape Convention with territorial planning also promotes communication among different sectors and stakeholders. Such a territorial vision for the conservation of heritage resources can effectively reduce the cost of climate change in terms of flood and landslide in Central Italy. Our analysis also shows that different municipalities have different levels of vulnerability to these risks based on their information and skills and institutional preparedness. One reason is that environmental organizations and public communication by scientists play a marginal role in terms of climate adaptation of cultural landscapes in Central Italy. Further research is required to understand how territorial planning can reduce organizational tensions in defining protection priorities as well as assessing stakeholder preparedness for necessary actions in terms of managing climate change and natural risks.

Author Contributions: Conceptualization, A.S.D., S.B.A.; methodology, A.S.D., S.B.A., and A.C.; writing—original draft preparation, A.S.D., S.B.A.; writing—review and editing, A.S.D., S.B.A., A.C., M.S., and G.D.L.; supervision, S.B.A., A.C., and G.D.L.; All authors have read and agree to the published version of the manuscript.

Funding: This paper was developed in the frame of Italy–USA Resilient Landscapes (RELAND), funded by the Ministry of Foreign Affairs and International Cooperation of Italy (MAECI), grant number of PGR06129.

Acknowledgments: The authors wish to express their special thanks to the US and Italian partners of the RELAND project for their precious collaboration.

Conflicts of Interest: The authors declare no conflict of interest.

References

1. Jones, M. The concept of cultural landscape: Discourse and narratives. In *Landscape Interfaces*; Springer: Berlin, Germany, 2003; pp. 21–51.
2. Perry, J. Climate Change Adaptation in Natural World Heritage Sites: A Triage Approach. *Climate* **2019**, *7*, 105. [CrossRef]
3. Collete, A. Climate Change and World Heritage: Report on Predicting andMmanaging the Impacts of Climate Change on World Heritage and Strategy to Assist States Parties to Implement Appropriate Management Responses. Available online: http://whc.unesco.org/document/6670 (accessed on 25 June 2019).
4. UNESCO World Heritage Center. *The Operational Guidelines for the Implementation of the World Heritage Convention*; UNESCO World Heritage Center: Paris, France, 2019.
5. Shirvani Dastgerdi, A.; De Luca, G. The Riddles of Historic Urban Quarters Inscription on the UNESCO World Heritage List. *Int. J. Archit. Res. ArchNet-IJAR* **2018**, *12*, 152–163. [CrossRef]
6. ICCROM European Cultural Heritage Summit—Sharing Heritage, Sharing Values. Available online: https://www.iccrom.org/news/european-cultural-heritage-summit-sharing-heritage-sharing-values (accessed on 2 September 2019).
7. Sargolini, M. Urban landscapes and nature in planning and spatial strategies. In *Nature Policies and Landscape Policies*; Springer: Berlin, Germany, 2015; pp. 299–306.
8. Morandi, F.; Niccolini, F.; Sargolini, M. *Parks and Territory: New Perspectives in Planning and Organization*; LISt Lab Laboratorio Internazioale Editoriale: Trento, Italy, 2012.
9. Melnick, R.Z.; Kerr, N.P.; Malinay, K.; Burry-Trice, O. *Climate Change and Cultural Landscapes: A Guide to Research, Planning, and Stewardship*; University of Oregon Press: Eugene, OR, USA, 2017. [CrossRef]
10. Colette, A. *Case Studies on Climate Change and World Heritage*; UNESCO World Heritage Center: Paris, France, 2007.
11. Sabbioni, C.; Cassar, M.; Brimblecombe, P.; Tidblad, J.; Kozlowski, R.; Drdácký, M.; Saiz-Jimenez, C.; Grøntoft, T.; Wainwright, I.; Ariño, X. Global climate change impact on built heritage and cultural landscapes. In Proceedings of the Proceedings of the International Conference on Heritage, Weathering and Conservation (HWC 2006), Madrid, Spain, 21–24 June 2006.
12. Fatorić, S.; Seekamp, E. Are cultural heritage and resources threatened by climate change? A systematic literature review. *Clim. Chang.* **2017**, *142*, 227–254. [CrossRef]
13. Bertolin, C. Preservation of Cultural Heritage and Resources Threatened by Climate Change. *Geosciences* **2019**, *9*, 250. [CrossRef]

14. Perry, J.; Falzon, C. *Climate Change Adaptation for Natural World Heritage Sites: A Practical Guide*; UNESCO World Heritage Center: Paris, France, 2014; Volume 37.

15. Hulme, M.; Jenkins, G.J.; Lu, X.; Turnpenny, J.R.; Mitchell, T.D.; Jones, R.G.; Lowe, J.; Murphy, J.M.; Hassell, D.; Boorman, P.; et al. Climate change scenarios for the United Kingdom: The UKCIP02 scientific report. Tyndall Centre for Climate Change Research, School of Environmental Sciences, University of East Anglia, Norwich, 2002.[Accedido 30 de Sept 2005]. *Nat. Clim. Chang.* **2015**, *5*.

16. Markham, A.; Osipova, E.; Lafrenz Samuels, K.; Caldas, A. World Heritage and Tourism in a Changing Climate. Available online: https://whc.unesco.org/document/139944 (accessed on 26 January 2020).

17. ICOMOS The Future of Our Pasts: Engaging Cultural Heritage in Climate Action. Available online: https://indd.adobe.com/view/a9a551e3-3b23-4127-99fd-a7a80d91a29e (accessed on 26 January 2020).

18. Holtz, D.; Markham, A.; Cell, K.; Ekwurzel, B. *National Landmarks at Risk: How Rising Seas, Floods, and Wildfires Are Threatening the United States' Most Cherished Historic Sites*; Union of Concerned Scientists: Cambridge, MA, USA, 2014.

19. IPCC Annex II: Glossary. In *Climate Change 2014: Synthesis Report. Contribution of Working Groups I, II and III to the Fifth Assessment Report of the Intergovernmental Panel on Climate Change*; Mach, K.J.; Planton, S.; von Stechow, C. (Eds.) IPCC Core Writing Team: Geneva, Switzerland, 2014.

20. Tsitsiragos, D. Climate Change Is a Threat—And an Opp Ortunity—For the Private Sector. Available online: https://www.worldbank.org/en/news/opinion/2016/01/13/climate-change-is-a-threat---and-an-opportunity---for-the-private-sector (accessed on 24 January 2020).

21. Georgeson, L.; Maslin, M.; Poessinouw, M.; Howard, S. Adaptation responses to climate change differ between global megacities. *Nat. Clim. Chang.* **2016**, *6*, 584. [CrossRef]

22. Dilling, L. Water security and adaptive capacity for climate: Learning lessons from drought decision making in U.S. urban contexts. *AGU Fall Meet. Abstr.* **2017**, *2017*, H13S-02.

23. Allen, M.R.; Dube, O.P.; Solecki, W.; Aragón-Durand, F.; Cramer, W.; Humphreys, S.; Kainuma, M.; Kala, J.; Mahowald, N.; Mulugetta, Y.; et al. Framing and Context. In *Global Warming of 1.5°C. An IPCC Special Report on the Impacts of Global Warming of 1.5°C above Pre-Industrial Levels and Related Global Greenhouse Gas Emission Pathways, in the Context of Strengthening the Global Response to the Threat of Climate Change*; Masson-Delmotte, V., Zhai, P., Pörtner, H.-O., Roberts, D., Skea, J., Shukla, P.R., Pirani, A., Moufouma-Okia, W., Péan, C., Pidcock, R., et al., Eds.; IPCC: Paris, France, 2018.

24. Siders, A.R. Adaptive capacity to climate change: A synthesis of concepts, methods, and findings in a fragmented field. *Wiley Interdiscip. Rev. Clim. Chang.* **2019**, *10*, e573. [CrossRef]

25. Shirvani Dastgerdi, A.; Sargolini, M.; Pierantoni, I. Climate Change Challenges to Existing Cultural Heritage Policy. *Sustainability* **2019**, *11*, 5227. [CrossRef]

26. Denton, F.; Wilbanks, T.J.; Abeysinghe, A.C.; Burton, I.; Gao, Q.; Lemos, M.C.; Masui, T.; O'Brien, K.L.; Warner, K.; Bhadwal, S.; et al. *Climate-Resilient Pathways: Adaptation, Mitigation, and Sustainable Development*; IPCC: Paris, France, 2015.

27. Sesana, E.; Gagnon, A.S.; Bertolin, C.; Hughes, J. Adapting cultural heritage to climate change risks: Perspectives of cultural heritage experts in europe. *Geosciences* **2018**, *8*, 305. [CrossRef]

28. Posey, J. The determinants of vulnerability and adaptive capacity at the municipal level: Evidence from floodplain management programs in the United States. *Glob. Environ. Chang.* **2009**, *19*, 482–493. [CrossRef]

29. Engle, N.L.; Lemos, M.C. Unpacking governance: Building adaptive capacity to climate change of river basins in Brazil. *Glob. Environ. Chang.* **2010**, *20*, 4–13. [CrossRef]

30. Smit, B.; Pilifosova, O. From adaptation to adaptive capacity and vulnerability reduction. In *Climate Change, Adaptive Capacity and Development*; IPCC: Paris, France, 2003; ISBN 9781860945816.

31. ESPON CLIMATE. *Climate Change and Territorial Effects on Regions and Local Economies*; ESPON CLIMATE: Luxembourg, 2010.

32. Sesana, E.; Bertolin, C.; Loli, A.; Gagnon, A.S.; Hughes, J.; Leissner, J. Increasing the Resilience of Cultural Heritage to Climate Change Through the Application of a Learning Strategy. *Int. Conf. Transdiscipl. Multispectral Model. Coop. Preserv. Cult. Herit.* **2018**, 402–423.

33. Yale program on Climate Change Communication. Available online: https://climatecommunication.yale.edu/ (accessed on 15 September 2019).

34. OCED. *Water and Climate Change Adaptation: Policies to Navigate UnchartedWaters*; OECD Studies on Water; OECDt: Geneva, Switzerland, 2013; ISBN 9789264200432.

35. Galeotti, M.; Roson, R. Economic Impacts of Climate Change in Italy and the Mediterranean: Updating the Evidence. *J. Sustain. Dev.* **2012**, *5*, 27–41. [CrossRef]

36. World Bank Italy. Available online: https://climateknowledgeportal.worldbank.org/country/italy (accessed on 11 September 2019).

37. Pinault, J.-L. Anthropogenic and Natural Radiative Forcing: Positive Feedbacks. *J. Mar. Sci. Eng.* **2018**, *6*, 146. [CrossRef]

38. Ministry for the Environment, Land and Sea. *Vulnerability Assessment, Climate Change Impacts and Adaptation Measures—Convention on Climate Change Italy;* Ministry for the Environment, Land and Sea: Rome, Italy, 2009.

39. Ministry of the Environment and Land Protection Flood Risk Management in Italy: Tools for the Hydrogeological Land Llanning. Available online: https://www.minambiente.it/sites/default/files/archivio/biblioteca/ds_flood_risk_management_vienna17_18_05_2006.pdf (accessed on 15 September 2019).

40. Peter Sommer Travels Italian Geography. Available online: https://www.petersommer.com/italy/geography (accessed on 26 January 2020).

41. World Heritage Centre World Heritage List Statistics. Available online: https://whc.unesco.org/en/list/stat/ (accessed on 1 October 2019).

42. World Heritage Centre Interactive Map. Available online: https://whc.unesco.org/en/interactive-map/ (accessed on 6 January 2020).

43. World Heritage Centre World Heritage List. Available online: https://whc.unesco.org/en/list (accessed on 8 September 2019).

44. UNESCO World Heritage Center Val d'Orcia. Available online: http://whc.unesco.org/en/list/1026 (accessed on 26 January 2020).

45. UNESCO World Heritage Centre Villa d'Este, Tivoli. Available online: http://whc.unesco.org/en/list/1025 (accessed on 26 January 2020).

46. Ministero per i Beni e le Attività Culturali Legislative Decree No. 42 of 22 January 2004- Code of the Cultural and Landscape Heritage. Available online: https://sherloc.unodc.org/res/cld/document/ita/code-of-the-cultural-and-landscape-heritage_html/it_cult_landscapeheritge2004_engtof.pdf (accessed on 26 January 2020).

47. UNESCO Convention Concerning the Protection of the World Cultural and Natural Heritage. Available online: https://whc.unesco.org/en/conventiontext/ (accessed on 26 January 2020).

48. Sauer, C.O. The Morphology of Landscape. *Univ. Calif. Publ. Geogr.* **1925**, *22*, 19–53.

49. World Heritage Centre. *Operational Guidelines for the Implementation of the World Heritage Convention;* UNESCO World Heritage Centre: Paris, France, 1999.

50. Council of Europe. *European Landscape Convention;* Council of Europe: Strasbourg, France, 2000.

51. Trusiani, E. Cultural Landscape. In *Landscape: Between Conservation and Transformation;* Biscotto, E., D'Astoli, S.B., Trusian, E., Eds.; Gangemi: Rome, Italy, 2013.

52. Shirvani Dastgerdi, A.; De Luca, G. Specifying the Significance of Historic Sites in Heritage Planning. *Conserv. Sci. Cult. Herit.* **2018**, *18*, 29–39.

53. Amoruso, G.; Salerno, R. *Cultural Landscape in Practice;* Springer: Berlin, Germany, 2019.

54. Sargolini, M. *Mountain Landscapes;* University of Camerino: Camerino, Italy, 2016; ISBN 9788898774388.

55. ISTAT Temperature and Precipitation in the Main Cities. Available online: https://www.istat.it/it/archivio/217402 (accessed on 2 January 2020).

56. Bartolini, G.; Morabito, M.; Crisci, A.; Grifoni, D.; Torrigiani, T.; Petralli, M.; Maracchi, G.; Orlandini, S. Recent trends in Tuscany (Italy) summer temperature and indices of extremes. *Int. J. Climatol. A J. R. Meteorol. Soc.* **2008**, *28*, 1751–1760. [CrossRef]

57. World Bank Climate Data. Available online: https://climateknowledgeportal.worldbank.org/country/italy/climate-data-historical (accessed on 1 September 2019).

58. Normativa nazionale Legge 3 agosto 1998, n. 267. Available online: http://www.edizionieuropee.it/LAW/HTML/4/zn17_03_163.html (accessed on 1 September 2019).

59. Sgobbi, A.; Carraro, C. *Climate Change Impacts and Adaptation Strategies in Italy: An Economic Assessment;* FEEM Fondazione Eni Enrico Mattei Reseach Paper; Fondazione Eni Enrico Mattei (FEEM): Milan, Italy, 2008.

60. ISTAT Hydrogeological Instability in Italy: Hazard and Risk Indicators. Available online: http://www.isprambiente.gov.it/it/pubblicazioni/rapporti/dissesto-idrogeologico-in-italia-pericolosita-e-indicatori-di-rischio-rapporto-2015 (accessed on 2 January 2020).

61. Perna, P.; Pierantoni, I.; Renzi, A.; Sargolini, M. *SUN LIFE: Strategy for managing the Natura 2000 network in Umbria*; LIST Laboratorio: Roma, Italy, 2018; ISBN 9788899854331.

62. Beltrame, L.; Bucchi, M.; Loner, E. Climate Change Communication in Italy. In *Oxford Research Encyclopedia of Climate Science*; Oxford University Press: Oxford, UK, 2017.

63. Asghar, S.; Alahakoon, D.; Churilov, L. A comprehensive conceptual model for disaster management. *J. Humanit. Assist.* **2006**, *1360*, 1–15.

64. Shirvani Dastgerdi, A.; Stimilli, F.; Pisano, C.; Sargolini, M.; De Luca, G. Heritage waste management: A possible paradigm shift in post-earthquake reconstruction in central Italy. *Cult. Herit. Manag. Sustain. Dev.* **2019**, 1–17. [CrossRef]

65. Harvey, D.C.; Perry, J. *The Future of Heritage as Climates Change: Loss, Adaptation and Creativity*; Routledge: London, UK, 2015; ISBN 9781317530121.

66. Magnaghi, A. *Il progetto locale: Verso la coscienza di luogo*; Bollati Boringhieri: Turin, Italy, 2013.

67. Gary, G.; Allred, S.B.; LoGiudice, E.; Chatrchyan, A.; Baglia, R.; Mayhew, T.; Olsen, D.; Wyman, M. Community Adaptation to Flooding in a Changing Climate: Municipal Officials' Actions Decision-Making and Barriers. *Res. Policy Br. Ser.* **2013**.

Article

Climate Change Adaptation in Natural World Heritage Sites: A Triage Approach

Jim Perry

Department of Fisheries, Wildlife and Conservation Biology, University of Minnesota, 2003 Upper Buford Circle, St Paul, MN 55108, USA; Jperry@umn.edu

Received: 14 August 2019; Accepted: 30 August 2019; Published: 2 September 2019

Abstract: Climate change is a certainty, but the degree and rate of change, as well as impacts of those changes are highly site-specific. Natural World Heritage sites represent a treasure to be managed and sustained for all humankind. Each World Heritage site is so designated on the basis of one or more Outstanding Universal Values. Because climate change impacts are site-specific, adaptation to sustain Universal Values also must be specific. As such, climate change adaptation is a wicked problem, with no clear action strategies available. Further, adaptation resources are limited at every site. Each site management team must decide which adaptations are appropriate investments. A triage approach guides that evaluation. Some impacts will be so large and/or uncertain that the highest probability of adaptation success comes from a series of uncertain actions that reduce investment risk. Others will be small, certain, comfortable and yet have low probable impact on the Universal Value. A triage approach guides the management team toward highest probable return on investment, involving stakeholders from the surrounding landscape, advancing engagement and communication, and increasing transparency and accountability.

Keywords: risk-based decisions; triage; protected areas; scenario planning

1. Introduction

Human society is now living in the Anthropocene [1–5], encountering far-reaching anthropogenic environmental changes. Climate change has become one of the few most critical challenges we face [6,7]. Accelerated, anthropogenic climate change threatens the capacity of the Earth's ecosystem to sustain human well-being. The global ecosystem has a limited capacity to meet the demands of a constantly increasing human population [8,9]. As such, climate change has become the quintessential crisis, attracting public and political attention [9]. Such attention varies among fear, denial, and mitigation/adaptation.

Although globally, climate change is a certainty [10,11], the rate and magnitude of specific changes are highly site-specific. Each management entity (i.e., those responsible for management of a given landscape) must develop and implement adaptation actions specific for the site itself. That specificity is demanding, expensive and yet critical when the site involved is unique and of global significance. Natural World Heritage (NWH) sites represent a pool of 252 landscape units, each of which is unique and of global significance (Figure 1). Sixteen of those 252 sites are classified as "In-Danger" by UNESCO, and two of the sixteen (i.e., Everglades and East Rennell) are specifically threatened by climate change. Others such as the Australian Wet Tropics are highly sensitive to small and almost certain changes in climate [12–14]. Climate change is a potential threat to each NWH [15], but especially so to ecosystems that are rich in biodiversity and endemism (e.g., the subtropical rainforests of central eastern Australia) [16]. More than a decade ago, the World Heritage Committee (WHC) instructed that all World Heritage (WH) site management plans assess the possible impact of climate change and prepare mitigation strategies [16–18]. That call has been repeated by several global groups, e.g., [19–21]. Meeting that need demands attention and resources that often are in limited supply.

Figure 1. Location of the 252 UNESCO Natural World Heritage (NWH) sites; the 16 sites in red are on UNESCO's "Sites In-Danger" list [22].

Climate change, and climate change adaptation make management of NWH sites and other large protected areas ever more challenging. Sustaining the attributes that make the site qualify as World Heritage becomes pursuit of a moving target. Past and current management strategies may not be adequate for WH sites and other large protected areas in adapting to a changing climate because the sites originally were developed and managed with the notion of static boundaries and with the aim of maintaining values present at the time of designation [23]. For example, Zaccarelli et al. [24] comment that conservation plans assume that biodiversity and human values both are static. Yet, increased awareness of the value of biodiversity has changed the priorities associated with it [25]. Similarly, Wu et al. [26] demonstrate the need for forward-looking conservation strategies that can adapt to the novel communities expected to represent future avifauna, a view clearly applicable to other taxa as well.

One of the attributes of climate change that is most widely accepted is that the frequency and magnitude of extreme events will increase before there is a demonstrable change in average conditions [27]. Geologic history shows that the Earth's climate has always been characterized by high variability, underlain by natural cycles. However, as humans increasingly have altered atmospheric gases, both the frequency and intensity of events have become more pronounced, with extremes of both temperature and rainfall commonplace [11,28,29]. For example, distinct Australian flood events with 1 in 100, 1 in 150 and 1 in 300 year return periods all were recorded in 2007 [11]. That change in climatic conditions necessitates a change in planning and management. But, because futures are so unpredictable the requirement is for an adaptable strategy, not a series of new practices.

2. Risks that Climate Change Poses to Natural World Heritage Sites

Each WH site has a series of attributes the cause it to meet the UNESCO criteria for WH. Those attributes represent the Outstanding Universal Value (OUV) that characterizes the site. Climate change may pose a wide range of stressors to the variables that constitute the OUV. Examples of such stressors include sea level rise, changes in temperature and precipitation, and altered hydrologic regimes. Cloud forests such as those in the mountainous regions of Western Australia and the Andes are especially sensitive to such changes [12–14,30,31]. Coral reefs such as those in Belize [32], Australia [33] and Indonesia [34] have been demonstrated to be highly responsive to climate changes. I developed a World Heritage Vulnerability Index (WHVI) for NWH sites, to identify the sites at greatest risk from climate change [35]. That index relied on nine variables, ranked all NWH sites (220 at the time), and could inform debate concerning the future management of key heritage assets [11]. However, that approach served as a comparison among sites, not as guidance for an individual site. In a second contribution [36], a colleague and I offered a range of site-specific, climate change adaptation actions within and around any given NWH site. Decisions and actions within and among NWH sites were placed in a theoretical context in a later paper [37]. The current contribution takes that work forward to offer decision-making guidance applicable to all NWH sites.

3. Triage as Guidance for Prioritization

Triage is a term that has been used in the medical community since the early 1900s to sort patients into categories: (1) deceased or untreatable; (2) critical; (3) stable; and (4) minor. This paper argues that decisions at the site-scale necessarily follow the logic of triage. That process for prioritizing investment of scarce resources is implicitly applied on a daily basis by managers and policymakers, but its inherent logic (i.e., specific triage application) is most often implicit [38]. For example, Reimann et al. [39] followed a triage approach to offer an index that allows ranking WH at risk from coastal hazards and found it successful. Krosby et al. [40] used large temperature gradients, high canopy cover, large relative width, low exposure to solar radiation, and low levels of human modification to calculate a riparian climate-corridor index that results in triage-based risk evaluations ranging in scale from local watersheds to the entire Pacific Northwestern US. But, by failing to be explicit and transparent, many decisions will necessarily be inefficient [41].

Disaster risk reduction (DRR) is a large and critically important field that attempts to guide society in preparing for and responding to natural disasters. Human society has always been subject to disaster, a history which logically would lead to advanced states of prediction and planning. In fact, however, society broadly, and many governments in particular appear to have short memories of past disasters and very low willingness to invest in disaster avoidance [42]. Science and policy have come to recognize that integrating disaster response and climate change adaptation is a necessity [43]. A wide range of frameworks has been developed for framing societal actions [44,45]. In spite of that breadth, the linkage between DRR and climate change response has yet to emerge. For example, at the time of this writing (21 August 2019), Web of Science reported 3474 papers that addressed "disaster risk reduction", but only six of those addressed climate change. Only one [46] addressed climate change and triage, and the context was medical response not preparedness or adaptation. In spite of the logical similarity between DRR preparedness and climate adaptation planning, the frameworks advanced for the former are not useful for site-specific or landscape-scale climate change adaptation planning for NWH.

Bottrill et al. [41] suggest that conservation triage can serve effectively for prioritizing actions as long as alternative actions are assessed relative to at least four parameters: values, biodiversity benefit, probability of success, and cost, necessarily combining those four in a mathematically rigorous fashion. Of course, understanding the probability of success requires a clear overview of the actual situation and clear future risk scenarios. Further, triage and transparency by themselves will not solve the problem. I have suggested that climate change adaptation at NWH sites is a wicked problem for a variety of reasons [37], including the fact that there is an ever-changing baseline. Wicked problems have a range of characteristics that challenge climate change decision-making: each manager is faced with limited resources, competing public interests, increasing and novel threats, changing political environments and demands from a diversity of stakeholders [23]. Many of those novel threats will be unknown (in frequency and magnitude), making it difficult to judge probable successes. In every NWH site, some climate change impacts will be unmanageable given available resources, and that will prevent site managers from meeting the goal of sustaining all values [23].

Philips [16] suggests that one of the principal challenges of adaptation is that there are no limits to the time and resources that could be absorbed and the absence of a limit makes it difficult to know how much to spend on adaptation. Further, there are widespread concerns about data reliability and the contested nature of information about climate change, resulting in a perceived lack of the necessary information to make fully informed decisions [16].

It is becoming increasingly clear that NWH sites and protected areas (PAs) are social-ecological systems [47]. As such, these resources are vulnerable to political change [48–50], economic fluctuations, and ecological variance [47] (Figure 2). Protected areas (including NWH sites) cannot be effectively managed based solely on ecological principles and the surrounding environment in which they are embedded [50]. Biophysical processes (e.g., landscape evolution, climate change) as well as sociopolitical dynamics drive questions like what specifically should be preserved, why and how? The most specific and practical guidance available for site-specific climate adaptation at individual NWH sites is the UNESCO Guide [36]. That guide and the paper offered here frame decisions as risk. Explicit triage logic guides decisions about such risks.

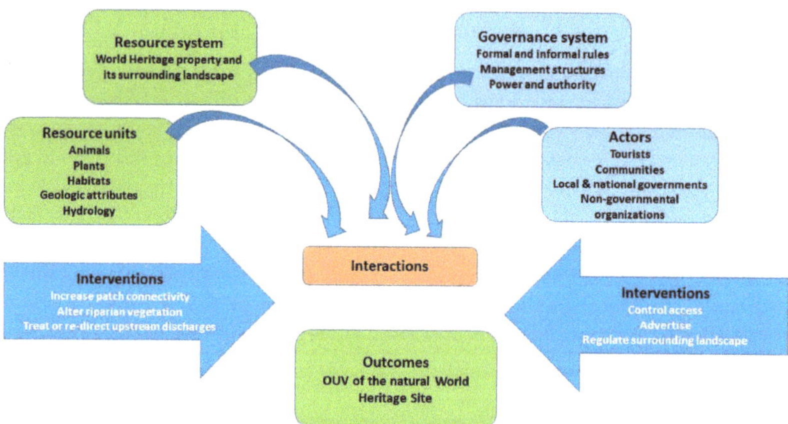

Figure 2. Biophysical and socioeconomic influences on the Outstanding Universal Value (OUV) of a NWH site, and examples of interventions as part of climate change adaptation. Modified from [47].

Every WH site has one or more OUVs. An OUV implies cultural and/or natural significance which is so exceptional that it transcends national boundaries and is " ... cultural and/or natural significance which is so exceptional as to transcend national boundaries and to be of common importance for present and future generations of all humanity." [22]. The reason for management to consider climate change adaptation is to ensure that the OUV of the WH site is sustained, which requires resilience in the face of climate change. That requires that we address at least two key biophysical factors for assessing the impact of climate change: (1) the character of the site within the natural landscape; and (2) the sensitivity and vulnerability of that natural landscape to process-driven, geomorphologic change [9]. But management decisions must consider more than that. Every decision must weigh the degree to which maintenance and intervention can be deemed cost-effective (based on the state party's definition) and the degree to which adaptation actions will be compatible with the aesthetic and other qualities that are part of the OUV [11].

It is appropriate that climate change adaptation at NWH sites focuses on the OUV. However, that focus cannot be blind to the natural processes of landscape change. Kabat et al. [51] offer three broad challenges that should guide such awareness: (1) conservation in a changing system requires a focus on the values and not the state of the system; (2) local and regional management should be nested in more coarse-scale governance structures; and (3) the management approach must be sufficiently adaptable that it is able to deal with uncertainty and nonlinearities.

4. Guidance for Triage-Based Decisions

A site manager, or the site management team must be prepared to take calculated risks and choose optimal solutions in the context of what is known and understood. A triage approach supports that. Triage guides allocation of conservation resources such that scarce resources are allocated to maximize persistence of features that will disappear without conservation action [37]. That is, based on climate predictions and scenarios of the future, some attributes of the site (and of the OUV) will be relatively unaffected, some will almost surely be altered in spite of management action, and some are likely to be responsive to (protected by) management action. Investing time and resources in the latter of those three maximizes the probability of a successful outcome (Figure 3). Assessing the vulnerability of a site's features against climate scenarios will help the site team assess the degree of risk climate change poses to the site's OUV. Scenarios incorporate interacting risks and uncertainties, advancing informed decisions [52]. This in turn, will enable the management team to prioritize responses against criteria, providing a basis for action that can be monitored and reviewed. By explicitly acknowledging that scenarios and decisions are based in triage, the manager will be able to evaluate tradeoffs and provide transparency of decision-making [39].

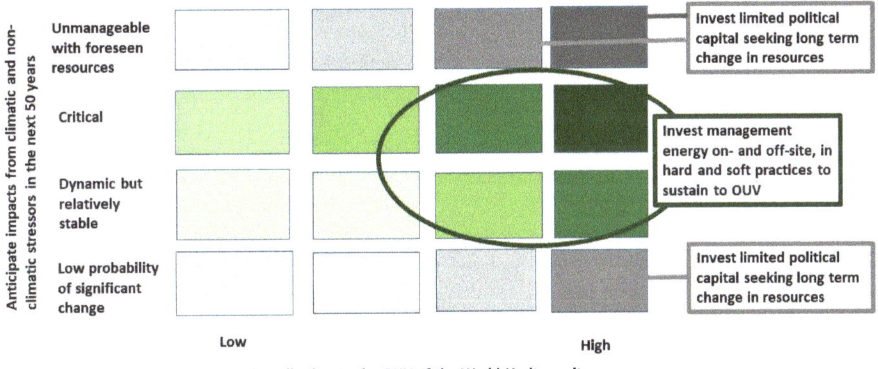

Figure 3. A risk triage for managing the OUV of a NWH site. Modified from [38]. For example, a strategy in the lower right (low risk, high value) might be to expand the boundaries of the site, a low probability effort that would require significant time and energy. Strategies in the upper right represent high value attributes that probably cannot be saved (e.g., eroding coastlines). This approach recognizes that valuable resources (i.e., upper right corner) will be lost without additional resources, but argues that long term sustainability of the OUV comes from investment in the right center. Examples of such investments might include wildlife corridors to connect patches, or no-take marine protected zones for reproduction.

While future climate conditions are very difficult to predict precisely, even rough predictions in the form of scenarios will help a manager think about the ways in which attributes of the OUV may be expected to respond to future climate conditions. This allows at least some form of risk analysis as the basis for designing an adaptation plan. Such a plan should provide a range of prioritized actions, both within and beyond the site itself. Those actions should include actions to be taken or proposed in the broader landscape. The most successful climate change adaptation strategies view the site as an element of a larger landscape and then address the OUV on-site in the context of off-site practices that influence the OUV [36].

Examples of such on- and off-site impacts abound. The Sundarbans is a massive wetland complex in India and Bangladesh. Some off-site impacts (e.g., sea level rise) are beyond the consideration of the management team. Others (e.g., deforestation in the headwaters, mining in adjacent properties) can be influenced by but not controlled by the management team. On-site practices such as hunting and shrimp culture can be directly controlled; others, such as on-site effects of flooding cannot [53]. However, traditional management has viewed the OUV of the site as the operational variable and the site boundaries as the spatial sphere of interest. A triage approach that considers off-site and on-site factors, and uses risk analysis logic to guide investments might, for example, guide the team toward mining and shrimp culture as the two most opportune investments, and expansion of site boundaries as one viable, long term strategy for moving forward.

The team should recognize that climate (and other anthropogenic) changes in the larger landscape will make future protected area management even more challenging than traditional management. Many traditional techniques will not be adequate for future conditions because NWH sites have been developed and managed in conceptual isolation, with the notion of static boundaries and the aim of maintaining current values [23,37]. It is common for managers and their teams to focus on lands under their control (or influence). For example, Fischman et al. [54] found that few protected area managers prescribed acting outside the refuge to address climate change impacts. They report that their finding is particularly surprising because protected area managers showed overall high rates of prescriptions for acting outside of refuge boundaries to address other problems, especially addressing water pollution, habitat loss, and invasive species. Yet protected areas (including NWH sites) are part of regional zones of social-ecological interactions. Management actions inside a site (e.g., ecological connectivity, water flows) impact, and are impacted by responses (e.g., development) outside site boundaries [55]. Actions outside site boundaries are often controlled by political, bureaucratic, administrative or ownership variables, but cannot be ignored by site managers. As Kabat et al. [51] and Jones et al. [56] suggest, global trends in the economy cause an increasing need to evaluate projects on more coarse spatial scales; I would add the need for expanded temporal scales.

Specifically with regard to temporal influences, the site team must think into the future. Managers are less able than ever to predict what will happen under future climate conditions, but a diverse team can frame scenarios and make educated guesses. Climate change adaptation requires analysis of the current situation in light of historical and projected changes, measuring the results of actions taken, revising them and trying again. Adaptive management is based on this cycle of analysis, application, evaluation and revision [41]. That forward looking approach may seem intuitive, but is not traditional. Many of the recommendations for improving climate change adaptation encourage managing for change and adopting landscape-scale strategies [23]. Global changes such as increased mobility of many species, and increases in human alteration of the landscape mean that NWH sites can no longer be managed as ecological islands, independent of the broader social-ecological system in which they are located. Resilience of a site requires an ability to adapt to changing social and ecological conditions over time, supporting persistence of populations, communities, and ecosystems inherent to the OUV [47].

Adaptation practices at a NWH site are implemented at different levels. Site-specific, lower-level actions can be implemented by the management team responsible for the site. However, each site is nested within a larger biophysical and socioeconomic landscape that strongly influences site conditions. Higher-level actions involving stakeholders such as the surrounding community, policy-makers or energy and water companies also need to be considered. Practices at the site level are generally less expensive and provide quicker responses but may have more limited impact on protecting the OUV [36]. Critically, however, each decision to implement a practice should be framed as a triage, recognizing that various actions have different probabilities of impact, and each action represents a tradeoff, an investment of resources that could be used elsewhere.

5. Sustaining an Evolving Outstanding Universal Value: Manage with Transparency and Accountability

In initiating a triage approach, the NWH site management team will initially focus on the OUV; if the protected area is not a WH site, there will be a similar logic involving identified attributes deserving protection. The OUV and its attributes provide a framework which by itself does not offer enough detail for guiding adaptive responses. Vulnerability, the probability of climate change damage to the attributes of the OUV, or to ecosystem resilience provides the risk estimate. Vulnerability assessments go beyond threat assessments in that they help the management team understand the capacity of the site to withstand or adapt to climate change (and other anthropogenic) impacts [54]. Climate change vulnerability specifically identifies the extent to which predicted climatic conditions are likely to cause a negative impact to the site's OUV. Expressions of vulnerability necessarily include cultural processes (e.g., socioeconomic influences, adaptation and mitigation actions, governance). The site team must consider those variables through engaging stakeholders in the surrounding landscape. As NWH site managers become more aware of climate impacts, the team must become more tightly connected with regional and national governance structures [51]. Inherent in that increase in scale and increase in inter-jurisdictional linkages will be the need to address and function under conditions of increased uncertainty [51]. The increasingly coarse spatial and temporal scale, the increased institutional complexity and increased uncertainty support the argument that this is a wicked problem [37] but also support the argument that a triage approach is necessary. The team should ensure that management goals are well communicated with all relevant stakeholders including those in the WH system (e.g., UNESCO, national government) and principal parties in the regional landscape. That transparency results in broader buy-in and wider acceptance of and support for decision-making under uncertainty. Decisions taken and actions performed should be monitored and evaluated such that the broader community interprets team behavior with a sense of accountability. No management team will be able to meet all expectations. Encouraging transparency and accountability provides the support necessary to take risky decisions.

6. Conclusions

A site manager, or the site management team for a NWH site must be prepared to take calculated risks and choose optimal solutions in the context of what is known and predicted about socioeconomic and biophysical conditions and future climates. Those decisions will be made under high uncertainty. Successful and sustainable management of a NWH will include both the site and the surrounding landscape, including the communities in that landscape. Decision-making under uncertainty will be most successful if the stakeholders have a strong sense that management is transparent and operates with a sense of accountability. Triage decision-making helps managers frame decisions in an explicit context that accepts the uncertainty involved, and advances both transparency and accountability.

Funding: This research was funded by the University of Minnesota.

Acknowledgments: I am grateful to Chiara Bertolin and three anonymous reviewers for helpful comments on earlier drafts of this manuscript.

Conflicts of Interest: The author declares no conflict of interest.

References

1. Ficek, R.E. Cattle, Capital, Colonization: Tracking Creatures of the Anthropocene in and out of Human Projects. *Curr. Anthr.* **2019**, *60*, S260–S271. [CrossRef]

2. Flitcroft, R.; Cooperman, M.S.; Harrison, I.J.; Juffe-Bignoli, D.; Boon, P.J. Theory and practice to conserve freshwater biodiversity in the Anthropocene. *Aquat. Conserv. Mar. Freshw. Ecosyst.* **2019**, *29*, 1013–1021. [CrossRef]

3. Harrison, H. Beyond "cultural" and "natural" heritage: Toward an ontological politics of heritage in the Anthropocene. *Herit. Soc.* **2015**, *8*, 24–42. [CrossRef]

4. Pagani-Núñez, E.; Liang, D.; He, C.; Zhou, X.; Luo, X.; Liu, Y.; Goodale, E. Niches in the Anthropocene: Passerine assemblages show niche expansion from natural to urban habitats. *Ecography* **2019**, *42*, 1360–1369. [CrossRef]

5. Sebastián-González, E.; Barbosa, J.M.; Pérez-García, J.M.; Morales-Reyes, Z.; Botella, F.; Olea, P.P.; Mateo-Tomás, P.; Moleón, M.; Hiraldo, F.; Arrondo, E.; et al. Scavenging in the Anthropocene: Human impact drives vertebrate scavenger species richness at a global scale. *Glob. Chang. Biol.* **2019**, *25*, 3005–3017. [CrossRef] [PubMed]

6. IPCC. AR5 Synthesis Report: Climate Change 2014. Available online: https://www.ipcc.ch/report/ar5/syr/ (accessed on 5 August 2019).

7. UNFCCC. Bali Climate Change Conference. 2007. Available online: https://unfccc.int/process-and-meetings/conferences/past-conferences/bali-climate-change-conference-december-2007/bali-climate-change-conference-december-2007-0 (accessed on 5 August 2019).

8. Mathevet, R.; Thompson, J.D.; Folke, C.; Chapin, F.S. Protected areas and their surrounding territory: Socioecological systems in the context of ecological solidarity. *Ecol. Appl.* **2016**, *26*, 5–16. [CrossRef] [PubMed]

9. Nkoana, E.M.; Waas, T.; Verbruggen, A.; Burman, C.J.; Huge, J. Analytic framework for assessing participation processes and outcomes of climate change adaptation tools. *Environ. Dev. Sustain.* **2017**, *19*, 1731–1760. [CrossRef]

10. Burns, W.C. Belt and Suspenders? The World Heritage Convention's Role in Confronting Climate Change. *Rev. Eur. Community Int. Environ. Law* **2009**, *18*, 148–163. [CrossRef]

11. Howard, A.J. Managing global heritage in the face of future climate change: The importance of understanding geological and geomorphological processes and hazards. *Int. J. Herit. Stud.* **2013**, *19*, 632–658. [CrossRef]

12. Hilbert, D.W.; Ostendorf, B.; Hopkins, M.S. Sensitivity of tropical forests to climate change in the humid tropics of north Queensland. *Austral Ecol.* **2001**, *26*, 590–603. [CrossRef]

13. Ostendorf, B.; Hilbert, D.W.; Hopkins, M.S. The effect of climate change on tropical rainforest vegetation pattern. *Ecol. Model.* **2001**, *145*, 211–224. [CrossRef]

14. Williams, S.E.; Bolitho, E.E.; Fox, S. Climate change in Australian tropical rainforests: An impending environmental catastrophe. *Proc. R. Soc. B Biol. Sci.* **2003**, *270*, 1887–1892. [CrossRef] [PubMed]

15. ABC News. World Heritage Sites 'Threatened by Climate Change'. 2008. Available online: https://www.abc.net.au/news/2008-06-21/world-heritage-sites-threatened-by-climate-change/2479250 (accessed on 21 August 2019).

16. Phillips, H. Adaptation to Climate Change at UK World Heritage Sites: Progress and Challenges. *Hist. Environ.* **2014**, *5*, 288–299. [CrossRef]

17. UNESCO. Protecting and Managing Effects of Climate Change on World Heritage. Report to the 30th World Heritage Committee Meeting, Vilnius. 2006. Available online: http://whc.unesco.org/document/6670%20 (accessed on 21 August 2019).

18. UNESCO. Policy Document on the Impacts of Climate Change on World Heritage on World Heritage Properties. 2008. Available online: https://whc.unesco.org/en/news/441 (accessed on 21 August 2019).

19. IUCN. Climate Change and World Heritage. Available online: https://www.iucn.org/theme/world-heritage/our-work/global-world-heritage-projects/climate-change-and-world-heritage (accessed on 20 August 2019).

20. ICOMOS. ND. Resolution on Climate Change and Cultural Heritage. Available online: https://www.usicomos.org/icomos-passes-resolution-on-climate-change-and-cultural-heritage/ (accessed on 20 August 2019).

21. EarthJustice. World Heritage and Climate Change: The Legal Responsibility of States to Reduce Their Contributions to Climate Change. A Great Barrier Reef Case Study. 2017. Available online: http://earthjustice.org/sites/default/files/files/World-Heritage-Climate-Change_March-2017.pdf (accessed on 20 August 2019).

22. NESCO World Heritage Centre. 2017 Operational Guidelines. Available online: https://whc.unesco.org/document/163852 (accessed on 30 August 2019).

23. Tanner-McAllister, S.L.; Rhodes, J.; Hockings, M. Managing for climate change on protected areas: An adaptive management decision making framework. *J. Environ. Manag.* **2017**, *204*, 510–518. [CrossRef] [PubMed]

24. Zaccarelli, N.; Riitters, K.; Petrosillo, I.; Zurlini, G. Indicating disturbance content and context for preserved areas. *Ecol. Indic.* **2008**, *8*, 841–853. [CrossRef]

25. Hardisty, A.R.; Michener, W.K.; Agosti, D.; García, E.A.; Bastin, L.; Belbin, L.; Bowser, A.; Buttigieg, P.L.; Canhos, D.A.; Egloff, W.; et al. The Bari Manifesto: An interoperability framework for essential biodiversity variables. *Ecol. Inform.* **2019**, *49*, 22–31. [CrossRef]

26. Wu, J.X.; Wilsey, C.B.; Taylor, L.; Schuurman, G.W. Projected avifaunal responses to climate change across the U.S. National Park System. *PLoS ONE* **2018**, *13*, e0190557. [CrossRef]

27. McNeely, J.A. Protected Areas, Biodiversity, and the Risks of Climate Change. *Adv. Nat. Technol. Hazards Res.* **2016**, *42*, 379–397.

28. Hulme, M.; Dessai, S. Negotiating future climates for public policy: A critical assessment of the development of climate scenarios for the UK. *Environ. Sci. Policy* **2008**, *11*, 54–70. [CrossRef]

29. Rehm, E.M.; Olivas, P.; Stroud, J.; Feeley, K.J. Losing your edge: Climate change and the conservation value of range-edge populations. *Ecol. Evol.* **2015**, *5*, 4315–4326. [CrossRef]

30. Villa, P.M.; Perez-Sanchez, A.J.; Nava, F.; Acevedo, A.; Cadenas, D.A. Local-scale Seasonality Shapes Anuran Community Abundance in a Cloud Forest of the Tropical Andes. *Zool. Stud.* **2019**, *58*, 17.

31. Fadrique, B.; Báez, S.; Duque, Á.; Malizia, A.; Blundo, C.; Carilla, J.; Osinaga-Acosta, O.; Malizia, L.; Silman, M.; Farfán-Ríos, W.; et al. Widespread but heterogeneous responses of Andean forests to climate change. *Nature* **2018**, *564*, 207–212. [CrossRef] [PubMed]

32. Bove, C.B.; Ries, J.B.; Davies, S.W.; Westfield, I.T.; Umbanhowar, J.; Castillo, K.D. Common Caribbean corals exhibit highly variable responses to future acidification and warming. *Proc. R. Soc. B: Biol. Sci.* **2019**, *286*, 20182840. [CrossRef] [PubMed]

33. Pisapia, C.; Burn, D.; Pratchett, M.S. Changes in the population and community structure of corals during recent disturbances (February 2016–October 2017) on Maldivian coral reefs. *Sci. Rep.* **2019**, *9*, 842. [CrossRef] [PubMed]

34. Putra, R.D.; Suhana, M.P.; Kurniawn, D.; Abrar, M.; Siringoringo, R.M.; Sari, N.W.P.; Irawan, H.; Prayetno, E.; Apriadi, T.; Suryanti, A. Detection of reef scale thermal stress with Aqua and Terra MODIS satellite for coral bleaching phenomena. *AIP Conf. Proc.* **2019**, *2094*, 020024.

35. Perry, J. World Heritage hot spots: A global model identifies the 16 natural heritage properties on the World Heritage List most at risk from climate change. *Int. J. Herit. Stud.* **2011**, *17*, 426–441. [CrossRef]

36. Perry, J.; Falzon, C. *Climate Change Adaptation for Natural World Heritage Sites: A Practical Guide*; World Heritage Papers. 37; UNESCO: Paris, France, 2014; 88p, Available online: http://whc.unesco.org/document/129276 (accessed on 5 August 2019).

37. Perry, J. Climate change adaptation in the world's best places: A wicked problem in need of immediate attention. *Landsc. Urban Plan.* **2015**, *133*, 1–11. [CrossRef]

38. Lawler, J.J. Climate Change Adaptation Strategies for Resource Management and Conservation Planning. *Ann. N. Y. Acad. Sci.* **2009**, *1162*, 79–98. [CrossRef]

39. Reimann, L.; Vafeidis, A.T.; Brown, S.; Hinkel, J.; Tol, R.S.J. Mediterranean UNESCO World Heritage at risk from coastal flooding and erosion due to sea-level rise. *Nat. Commun.* **2018**, *9*, 4161. [CrossRef]

40. Krosby, M.; Theobald, D.M.; Norheim, R.; McRae, B.H. Identifying riparian climate corridors to inform climate adaptation planning. *PLoS ONE* **2018**, *13*, e0205156. [CrossRef]

41. Bottrill, M.C.; Joseph, L.N.; Carwardine, J.; Bode, M.; Cook, C.; Game, E.T.; Grantham, H.; Kark, S.; Linke, S.; McDonald-Madden, E.; et al. Is conservation triage just smart decision making? *Trends Ecol. Evol.* **2008**, *23*, 649–654. [CrossRef] [PubMed]

42. Taboroff, J. Cultural heritage and natural disasters: Incentives for risk management and mitigation. In *Managing Disaster Risk in Emerging Economies*; World Bank: New York, NY, USA, 2000; Volume 2, pp. 71–79.

43. Djalante, R.; Thomalla, F. Disaster risk reduction and climate adaptation in Indonesia-institutional challenges and opportunities for integration. *Int. J. Disaster Resil. Built Environ.* **2012**, *3*, 169–180.

44. O'Brien, G.; O'Keefe, P.; Jayawickrama, J.; Jigyasu, R. Developing a model for building resilience to climate risks for cultural heritage. *J. Cult. Herit. Manag. Sustain. Dev.* **2015**, *5*, 99–114. [CrossRef]

45. Wei, J.; Zhao, Y.; Xu, H.; Yu, H. A framework for selecting indicators to assess the sustainable development of the natural heritage site. *J. Mt. Sci.* **2007**, *4*, 321–330. [CrossRef]

46. Rumsey, M.; Fletcher, S.M.; Thiessen, J.; Gero, A.; Kuruppu, N.; Daly, J.; Buchan, J.; Willetts, J. A qualitative examination of the health workforce needs during climate change disaster response in Pacific Island Countries. *Hum. Resour. Health* **2014**, *12*, 9. [CrossRef] [PubMed]

47. Cumming, G.S.; Allen, C.R.; Ban, N.C.; Biggs, D.; Biggs, H.C.; Cumming, D.H.M.; De Vos, A.; Epstein, G.; Etienne, M.; Maciejewski, K.; et al. Understanding protected area resilience: A multi-scale, social-ecological approach. *Ecol. Appl.* **2015**, *25*, 299–319. [CrossRef] [PubMed]

48. Agrawal, A.; Gupta, K. Decentralization and Participation: The Governance of Common Pool Resources in Nepal's Terai. *World Dev.* **2005**, *33*, 1101–1114. [CrossRef]

49. Clement, S.; Moore, S.A.; Lockwood, M.; Mitchell, M. Using insights from pragmatism to develop reforms that strengthen institutional competence for conserving biodiversity. *Policy Sci.* **2015**, *48*, 463–489. [CrossRef]

50. Beeton, T.A.; McNeeley, S.M.; Miller, B.W.; Ojima, D.S. Grounding simulation models with qualitative case studies: Toward a holistic framework to make climate science usable for US public land management. *Clim. Risk Manag.* **2019**, *23*, 50–66. [CrossRef]

51. Kabat, P.; Bazelmans, J.; Van Dijk, J.; Herman, P.M.; Van Oijen, T.; Pejrup, M.; Reise, K.; Speelman, H.; Wolff, W.J. The Wadden Sea Region: Towards a science for sustainable development. *Ocean Coast. Manag.* **2012**, *68*, 4–17. [CrossRef]

52. Ojoyi, M.; Mutanga, O.; Kahinda, J.M.; Odindi, J.; Abdel-Rahman, E.; Kahinda, J.-M.M. Scenario-based approach in dealing with climate change impacts in Central Tanzania. *Futures* **2017**, *85*, 30–41. [CrossRef]

53. Hassan, K.; Higham, J.; Wooliscroft, B.; Hopkins, D. Climate change and world heritage: A cross-border analysis of the Sundarbans (Bangladesh-India). *J. Policy Res. Tour. Leis. Events* **2019**, *11*, 196–219. [CrossRef]

54. Fischman, R.L.; Meretsky, V.J.; Babko, A.; Kennedy, M.; Liu, L.; Robinson, M.; Wambugu, S. Planning for Adaptation to Climate Change: Lessons from the US National Wildlife Refuge System. *BioScience* **2014**, *64*, 993–1005. [CrossRef]

55. Hansen, A.J.; Davis, C.R.; Piekielek, N.; Gropss, J.; Thebold, D.M.; Goetz, S.; Melton, F.; DeFries, R. Delineating the Ecosystems Containing Protected Areas for Monitoring and Management BioScience 2011, 61, 363–273. *BioScience* **2011**, *61*, 363–373. [CrossRef]

56. Jones, K.R.; Watson, J.E.M.; Possingham, H.P.; Klein, C.J. Incorporating climate change into spatial conservation prioritisation: A review. *Biol. Conserv.* **2019**, *194*, 121–130. [CrossRef]

Article

Monitoring Climate Change in World Heritage Properties: Evaluating Landscape-Based Approach in the State of Conservation System

Paloma Guzman [1,*], Sandra Fatorić [2] and Maya Ishizawa [3]

[1] Norwegian Institute for Cultural Heritage Research (NIKU), 0155 Oslo, Norway
[2] Faculty of Architecture and the Built Environment, Delft University of Technology (TU Delft), 2628 BL Delft, The Netherlands; s.fatoric@tudelft.nl
[3] Department of World Heritage Studies, University of Tsukuba, Tsukuba 305-8577, Japan; maya@heritage.tsukuba.ac.jp
* Correspondence: paloma.guzman@niku.no

Received: 31 January 2020; Accepted: 4 March 2020; Published: 6 March 2020

Abstract: Climate change is increasingly being recognized as a threat to natural and cultural World Heritage (WH) sites worldwide. Through its interaction with other stressors, climate change accelerates existing risks while also creating new obstacles. A more considerable focus is needed in both research and practice to explore proactive measures for combatting this issue (e.g., mitigation and actions prior to impacts occurring). World Heritage values in climate change decision-making processes is an important factor in this regard. This paper explores a discussion of climate change within the WH monitoring system. It offers an overview of practice based on the extent to which WH properties (natural, mixed and cultural) implement landscape-based approaches alongside the conservation and management of their outstanding universal value within the context of climate uncertainty and environmental change. Landscape approaches are gaining importance in the WH conservation system, where they aim to provide concepts and tools for managing heritage toward sustainable practices. This research analyses the state of conservation reports and provides an overview of practice across time, categories and geographical regions. Based on a theoretical approach, empirical analyses identify four landscape principles that are increasingly shaping the debate around climate change issues in WH properties. Although these are highly relevant to advancing much-needed collaboration among scientific disciplines and governance sectors, we argue that further understanding is required on the transformational process of heritage values, as well as on the nature–culture relationship, in order to underpin heritage as a source for local resilience and climate mitigation.

Keywords: climate change; world heritage; landscape approach; nature culture divide; integrative heritage management; monitoring; state of conservation reports

---

1. Introduction

In the coming decades, climate change risks to various socioeconomic and natural systems are expected to increase due to projected anthropogenic climate changes, demographic development, and land-use changes [1]. These changes are increasingly being recognised as threat multipliers for global natural and cultural heritage via their interaction with other stressors, thereby accelerating existing risks and creating new ones [1–3]. An initial report on climate change and World Heritage (WH) properties (The term 'World Heritage properties' is used as indicated in the documents of the W System (Article 3 of the WH Convention). [4] stated that natural heritage may be jeopardised, whereas cultural properties (in addition to physical threats) will experience social and cultural impacts. In 2008,

The United Nations Educational, Scientific and Cultural Organisation (UNESCO) released the 'Policy Document on the Impacts of Climate Change on World Heritage Properties'. The document highlights that 'communities [have changed] the way they live, work, worship and socialize in buildings, sites and landscapes, [some] migrating and abandoning their built heritage'; the document further suggests that 'climate change will be considered in all aspects of nominating, managing, monitoring and reporting on the status of these properties' [5]. Accordingly, it follows recognising that in addition to 'the overarching objective of safeguarding the outstanding universal values of WH properties', these properties must 'serve as laboratories where monitoring, mitigation and adaptation processes can be applied, tested and improved' [5]. However, recent studies [3,6,7] found that heritage conservation practice has primarily focused on reactive measures that are taken after climate change impacts occur; these impacts include damage or loss to heritage properties [6,8,9].

The primary concern of WH decision makers and policymakers is maintaining the physical characteristics of heritage assets, through which their outstanding universal value (OUV) is conveyed [10,11]. This refers to the exceptional cultural or natural significance of WH properties [12]. Two aspects that warrant debate are embedded in the operationalisation of such requirements, which presents challenges to WH properties in terms of addressing climate change. The first is that conservation of WH values are often operationalised through static values and physical delimitations related to developing and managing properties [13–15]. The second consideration is that heritage values are categorised as natural and cultural, and their definition and monitoring are undertaken by different disciplinary entities [16,17]. Moreover, the fact that climate action related to WH has predominantly focused on risk and vulnerability assessments of their OUV [18] is linked to State party efforts that seek to avoid their properties being included in the list of WH in danger or even lose their designation as WH by being removed from the WH List [10]. The interest between State Parties to secure at any cost the positioning of their properties on the WH List, and the WH Centre and advisory bodies' efforts to promote transparent and ethical conservation practices, are increasingly clashing, rendering the WH Committee (WHCOM) a highly politicised arena [19]. This situation has not only damaged the credibility of the WHCOM but has also undermined the capacity of the WH monitoring system to provide fully transparent information about conservation processes [20].

The WH Centre and advisory bodies have made concrete efforts to advance systemic conceptualisations of heritage alongside sustainable conservation practices. Examples include recommendations and guidelines for the implementation of landscape approaches for managing the evolving character of properties' OUV [16,21], the integration of sustainable development perspectives to the processes of the World Heritage Convention (WHC) [22], as well as recent efforts to engage WH conservation in climate action [3,23]. Yet more consideration is needed in both research and practice to assess the efficiency of actions taken, and to explore or implement proactive measures that include WH values and aspects of climate change decision-making processes (e.g., designing and/or implementing efficient climate adaptation actions for WH properties) before impacts occur.

To address this gap, this paper recognises climate change as an adaptive challenge fundamentally linked to beliefs, values, and worldviews, as well as to power, politics, identity, and interests [24]. We consider a landscape-based approach to heritage conservation as a crucial strategy for overcoming the climate change challenge and biological and cultural diversity losses [16,25,26]. Using these concepts as a theoretical basis, this research explores how climate change is currently being discussed within the WH monitoring system. It offers a state of practice in relation to how WH properties (natural, mixed, and cultural) are implementing landscape-based approaches to the conservation and management of their OUV within the context of climate uncertainty and environmental change. To do this, we selected State of Conservation (SOC) reports to provide an overview of practice across time, categories, and geographical regions. SOC reports have largely been discussed in the literature as proof of the politicised conformation of the WH List [20,27] and have been criticised for lacking reliable measures to ensure the effective conservation of WH properties over time [28]. These reports have also proven to be a key source for providing valuable information on trends of conservation practices and on factors affecting the conservation of properties [29]. The empirical analyses presented herein identifies four landscape principles based on theory that are increasingly shaping the debate around climate change issues regarding WH properties. Although these are highly relevant for advancing much-needed collaboration among scientific disciplines and governance sectors, we argue that deeper understanding is required about the transformational process of heritage values and of the nature–culture relationship.

2. Landscape-Based Conservation and Monitoring Under Climate Change

The theoretical paradigm that leads conservation practice within the WH framework is based on a dichotomy [30,31], according to the naturalist ontology that influenced Western modern science, as it draws a distinction between humans and their natural environment [32,33]. This is exemplified by a distinction between natural and cultural heritage (defined in Articles 1 and 2 of the 1972 WH Convention) and the establishment of their respective advisory bodies. These include the International Union for Conservation of Nature (IUCN) for natural properties and the International Council of Monuments and Sites (ICOMOS) and the International Centre for the Study of the Preservation and Restoration of Cultural Property (ICCROM) for cultural properties. This division often causes clashes between the benefits of conserving natural values and those associated with cultural or social values, leading to competing priorities and conflicting interests in decision-making processes [34]. This is exacerbated to the extent that the conservation of socially constructed natural and cultural heritage values may become disassociated from their environmental settings and (climatic) processes [24,25,35,36].

There is growing consensus in the scientific literature [37–39] regarding the value of a landscape-based approach to conservation, which can help to bridge the conceptual gap between nature and culture. From this integrative perspective, heritage places include people, their built environments and practices, local ecosystems and other ecological processes. Although landscape-based approaches originated in the field of nature conservation [40,41], over the past decade, the cultural sector has also made efforts for its implementation [42–44]. Furthermore, landscape-based conservation practices are recognised for their potential to connect policy and practice at multiple interacting spatial levels via the implementation of adaptive and integrated management systems [38]. Moreover, landscape-based approaches are being consolidated as integrated solutions for addressing sustainability challenges, and to contribute to the fulfilment of the Sustainable Development Goals (SGD) of Agenda 2030 [25,45]. In the following subsections, we present common principles of landscape approaches discussed in the fields of natural and cultural conservation. These serve as analytical dimensions of the landscape-based approach and help to organise the primary findings presented in section three.

2.1. Landscape Scale and Governance

Landscape scale is acknowledged as a field in which various entities, including humans, interact according to physical, biological, and social rules that determine their relationships [46–48]. Here, the upscaling of conservation actions (generally targeting individual species or monuments) to an entire landscape helps to understand how objects of conservation relate to one another and how they are part of economic, environmental, social and cultural change processes [26,49,50]. In this way, landscape scale represents the wider physical/spatial context in which heritage resources are located, encompassing all possible interactions within a system [51], including the urban context [29]. Hence, conservation that employs a landscape-based approach requires managing the co-benefits and trade-offs between a wide range of stakeholders. In this way, conservation becomes a process of negotiation, decision making and re-evaluation that is expected to be informed by science but shaped by human values and preferences [46,49].

A shared conservation vision requires a governance structure for the coordination of different levels of organisation, assessment and implementation of measures. Thus, landscape governance involves the institutions, organisations, and mechanisms by which communities currently govern their relationship with the natural environment and global bio-geophysical systems [52]. The processes of institutionalising climate change approaches that facilitate effective internal networks toward collaboration and the resolution of complex objectives are often generally poorly understood in heritage management that deals with climate change, but also in climate change governance [34,53,54].

2.2. Evidence-Based Decision Making

Evidence-based decision making is a critical aspect for the normative of landscape [46] and adaptive management [55]. Mapping, assessment, and monitoring tools are expected to facilitate research activities for the description of socioecological systems. Based on this evidence, stakeholders can improve their capacity to judge and respond to manage heritage resources in transforming environments [56–58]. Consequently, monitoring using a landscape-based approach can provide a basis for revising and improving management and conservation practices by considering the complexity of human institutions and behaviours in the context of environmentally driven change [46,59]. In practice, this requires transdisciplinary and collaborative approaches that are able to validate different data levels and sources [29,60]. To articulate the evidence needed and produced by different stakeholders into individual value systems, a shared conservation vision must be agreed on alongside the coordination of activities for implementing relevant measures [61,62]. This requires all stakeholders to consider the outcomes of a shared conservation vision from a more holistic perspective [63,64].

2.3. Adaptation Measures and the Resilience of Natural and Cultural Heritage

According to the Intergovernmental Panel on Climate Change [1], the climate change adaptation of WH properties involves a decision-making process that seeks to reduce the vulnerability of WH areas to climate change and creating WH resilience to current and future climate change impacts [1]. Adaptation processes must consider the design and implementation of measures that can increase the resilience of landscape values and attributes [46,50] while ensuring that cultural and/or natural significance is not adversely affected or lost [65]. Plieninger et al. [14] discuss how landscape governance has been dominated by a conservation approach in which designated areas with special features should not change. Consequently, some studies have highlighted the importance of developing and operationalising landscape-based management frameworks to enable the safeguarding of heritage values and attributes while at the same time allowing for change [49,66]. However, discussion on the thresholds of 'acceptable change' remains at an early stage, particularly on the topics of climate change mitigation and adaptation fields [65,67–69]; a larger focus has been on socioeconomic development and its impacts on WH properties [70–72].

To sustain landscape processes and associated heritage values long-term, landscape resilience (the capacity to absorb and recover from impacts) represents a vital aspect of climate change decision-making processes [1,3]. The key elements in building heritage resilience include social/adaptive learning and the monitoring of changes [73,74]. Monitoring is a valuable tool for generating data and knowledge that can serve as a basis for informed decision making in managing WH properties within a changing climate environment [68,75].

3. Methodology

3.1. Analysis Sources

This study focused on using SOC reports and their qualitative data on WH management and conservation-related issues. The SOC reports, elaborated on a yearly basis, provided a brief analysis of the conservation threats to properties that both the WH Centre and advisory bodies considered as having high relevance for discussion by the WHCOM. These reports are requested when 'the values for which a property was inscribed on the WH List appear to be significantly threatened by either existing processes, or by potential processes with a high likelihood of taking place' [28] (e.g., climate change, development). Despite lacking quantitative information and analysis, SOC reports are the source of more objective discussions on WH issues such as the range of impacts on OUVs [12]. In their text format, reports include the viewpoints of the State party, advisory bodies, and the WH Centre in the form of concerns and recommendations that may differ from those of the national authorities or heritage managers [76]. Although SOC reports identify climate change and severe weather events among a list of 14 factors (https://whc.unesco.org/en/factors/) affecting the conservation of WH properties, the system does not systematically correlate extreme weather events (e.g., hurricane, drought) with climate change. Additionally, UNESCO's SOC database is not designed to provide systematic insights about the interactions between heritage management and the factors that both positively and negatively impact the conservation of WH properties. Nonetheless, it has proved valuable for the collation of information from numerous individual practices. Thus, SOC reports can be used to systematically assess patterns linked to the identification of threats to WH areas, management deficiencies, and conservation needs and developments [76]. For example, it has enabled the monitoring and quantification of the qualitative aspects of factors affecting the conservation of cultural and mixed heritage properties across time and geographical location [29].

3.2. Data Collection

In this study, we focused on 'climate change' as a key term. The SOC reports were downloaded from the UNESCO website and gathered in PDF format in order to conduct an advanced search in Acrobat Reader DC v.2020 using 'climate' + 'change' as keywords. Through directed content analysis, we sought to explore insights on the situational and management context in which climate change was noted in SOC monitoring exercises. The aim was to understand commonalities and differences between the challenges and/or opportunities discussed by conservation practice in all categories of WH properties when confronted by climate change as a global phenomenon of systemic complexity. We selected reports online from 2000 up to 2019.

3.3. Data Analysis

A coding system was applied to classify mentions of 'climate change' based on two pre-coding dimensions and two post-coding dimensions (Table 1). First, pre-coding identified mentions using the set of keywords 'climate change' and extracted all mentions found for analysis. The second pre-coding dimensions were SOC reports' general information, which was extracted and compiled in a Microsoft Excel spreadsheet in order to identify mentions according to the year of the SOC report, WH name, heritage category (i.e., natural, cultural, or mixed) and geographical location (country and region). The post-coding dimensions served to structure analysis involving advances in theory and practice. First, post-coding dimension three, context of references, identified the standard situational context in which mentions could be discussed, based on World Heritage Convention (WHC) protocols when elaborating on SOC reports. These included three viewpoints according to which debates on conservation issues revolved: (1) expression of concerns (C) by the WH Centre or advisory bodies on unresolved issues or possible negative outcomes of past and future measures; (2) recommendations (R) by both the WH Centre and advisory bodies to State Parties on prioritising actions; (3) implementations (I), which concerned the WHC's acknowledgement of efforts made by State Parties for dealing with a given threat or other conservation issues identified in previous WHC meetings. Viewpoints varied in order of appearance within reports. The second post-coding dimension, landscape-based conservation principles, comprised the operational definitions of landscape approaches identified through a taxonomic analysis of mentions. Four typologies were defined to formulate an initial framework for identifying landscape approaches related to climate change in WH conservation practices: (1) landscape scale, which refers to the upscaling of conservation activities in the wider context in which WH properties are located; (2) evidence of climate change impacts on OUV (lack or availability) to inform heritage management; (3) landscape governance, which includes the lack or existence of coherent planning and governance tools in collaboration with other sectors and actors (policies, management plans, and strategic actions); (4) adaptation measures addressing one of or the cumulative impacts of other identified factors that are exacerbated by climate change.

Table 1. Coding dimensions and definitions.

	Coding Dimension		**Coding Items and Definitions**
Pre-Coding	1. Frequency of mentions within SOC reports	1.	Climate + change (keywords)
	2. SOC reports' general information	1. 2. 3.	Year of SOC report Name of the property/category (natural, cultural, mixed) Geographical location
Post-Coding	3. Context of mentions	1. 2. 3.	(C) Expression of concern (R) Recommendation to implement measures (I) Implementations (identification of implemented measures)
	4. Landscape-based conservation principles	1. 2. 3. 4.	Landscape scale (the upscaling of conservation activities) Evidence of climate change impacts on OUV (lack or availability of monitoring and documentation tools to inform management actions) Landscape governance (lack or availability of coherent planning and management tools addressing local climate change and WH conservation) Adaptation measures (lack or availability of systemic measures addressing the impacts of climate change factors)

Collected qualitative data (mentions) were entered into an Excel spreadsheet and analysed using content analysis (Creswell, 2014). We calculated the frequency of each pre-coding and post-coding item, i.e., the number of times an item was coded across SOC reports. Next, relationships between pre and post-coding items were analysed based on their occurrence and following the hierarchical order of coding dimensions. The Excel COUNTIF function was used to count the number of extracted texts that met the criteria of coding dimensions. This allowed for interpretation of the meaning of mentions in the theoretical context of landscape conservation and the WH system. The key aspects of our analysis, which structured our results presented in the next section, were the following: (1) the complexity of discussion around climate change based on WH categories, regions (pre-coding dimensions one and two in Table 1) through the period 2000–2019; (2) the discussion of climate change drivers within the WH system (post-coding dimensions three and four in Table 1); (3) trends related to how WH conservation practice employed principles of landscape approaches to respond to climate change (relationships among all coding dimensions).

4. Results

4.1. Pre-Coding Analysis: Frequency of Mentions to 'Climate Change' in State of Conservation Reports

This section presents the results of the two pre-coding dimensions, frequency of mentions within SOC reports and SOC reports' general information (dimensions one and two in Table 1) and the correlation between their related items.

Although climate change risks to WH areas are acknowledged in the list of factors threatening properties (https://whc.unesco.org/en/factors/), our analysis showed that the term 'climate change' by itself was mentioned only 103 times within SOC reports in the period 2000–2019 (Figure 1). Mentions of climate change began appearing in 2000 and although these were not constant increased over time. The year 2017 had the highest number of mentions (n = 16), whereas 2002 and 2004 indicated no mentions. Mentions were observed referring to climate change as a general factor affecting the conservation of properties and rarely correlated to specific natural phenomena. The slow but increasing number of mentions across reports suggests that at the local level, climate change is still in an early identification stage within the WH system.

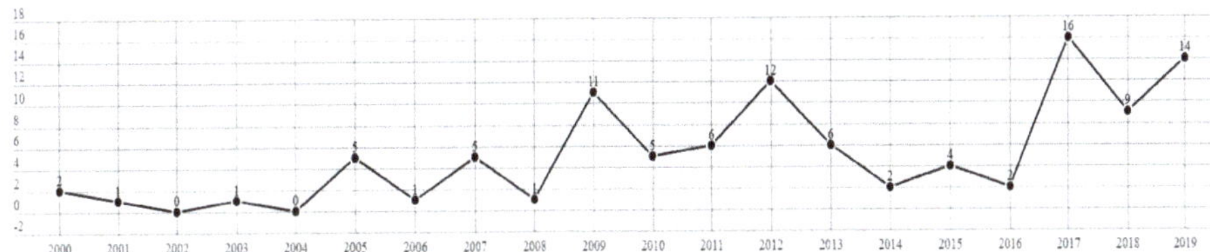

Figure 1. Timeline of total mentions (n = 103) of 'climate change' between 2000 and 2019.

Currently, 1121 properties are included in the WH List. Among these, 77.5% (869) are cultural properties, 19% (213) are natural properties, and 3.5% (39) are mixed properties. Only 574 of these properties have been discussed to date by the WHC and reported on in SOC reports. Similar to the distribution of categories on the WH List, cultural properties led the discussion on SOC reports on all negative impacts as it concerned the conservation of WH properties (68.5%, n = 393), followed by natural properties at 27% (n = 153) and 5% (n = 28) for mixed properties. This study found that the 103 mentions to climate change were discussed as having an impact on 64 properties, corresponding to 6% of all WH properties on the list and 11% of properties discussed in SOC reports. However, the analysis found that natural properties took the lead in the debate on climate change impacts on OUV. Approximately 80% of the total mentions of climate change were made by natural properties, followed by cultural properties at 14%. Mixed properties discussed climate change the least (6%) (see Figure 2).

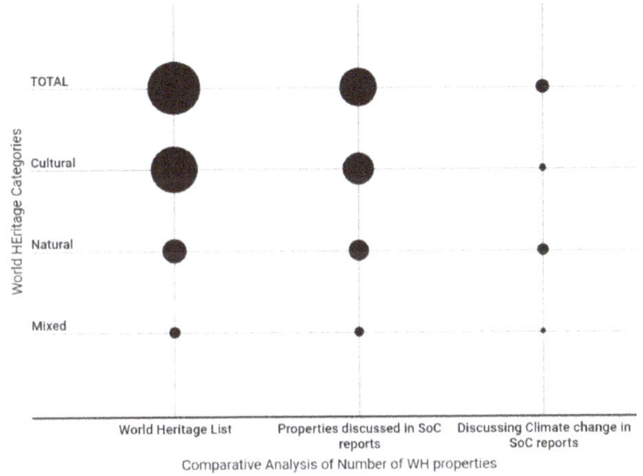

Figure 2. Comparison between total World Heritage (WH) properties and properties in State of Conservation reports discussing climate change.

Figure 3 shows the distribution of WH properties discussing climate change, as well as the percentage of mentions of climate change issues per UNESCO geographical region. The analysis of the geographical distribution of mentions to climate change showed that Asia and the Pacific (APA) reported on climate change the most, followed by Europe (EUR), Africa (AFR), Latin America, and the Caribbean (LAC) and Arab States (ARA). However, the number of WH properties being affected by climate change showed a slight change in regional order, where APA and AFR accounted for a major number (17 properties), followed by EUR with 16 properties, LAC with 11 and, finally, ARA with only three properties.

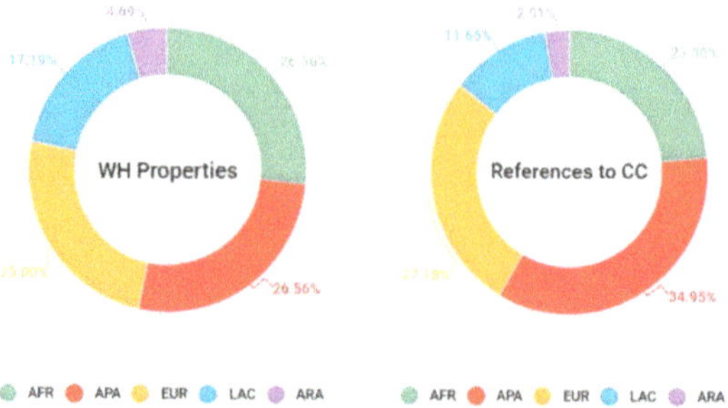

Figure 3. Distribution of WH properties discussing climate change (in %) and total references to climate change (cc) per The United Nations Educational, Scientific and Cultural Organisation (UNESCO) geographical area.

4.2. Post-Coding Analysis: The Context of Mentions and Landscape-Based Conservation Principles

This section discusses the results of summative post-coding dimensions three and four in Table 1. Relationships between the relevant years of SOC reports (dimension one, item one in Table 1) were employed to present the evolution of post-coding items over time. Post-coding dimension three, contexts of mentions, identified expression of concerns as the predominant context in which climate change was discussed, with 47.6% of all mentions. The second type of mentions identified the implementation of actions and measures (31%) presented by State Parties to the WH Centre. Rehabilitation activities and maintenance works, including concrete descriptions of actions such as the extension of boundaries and other adaptive interventions, were the most frequently mentioned measures. The final group corresponded to recommendations (21.4%) made by the WH Centre and advisory bodies to State Parties during the WHCOM; these commonly focused on further steps for management and conservation improvements. An overview of the evolution of context of mentions is shown in Figure 4. Specific examples across WH categories are discussed and illustrated in the following section.

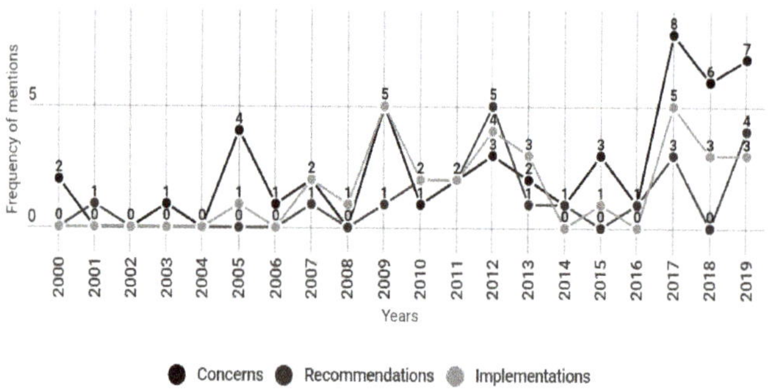

Figure 4. Mentions differentiated according to context (coding dimension 3).

Results from the post-coding analysis corresponding to dimension four in Table 1, landscape-based conservation principles, revealed the most pressing issue as a lack of evidence of climate change impacts on OUV (37.4% of all mentions) (Figure 5). This principle highlights the role of monitoring and documentation tools for informing management actions and conservation measures. The second principle, adaptation measures, was observed in roughly one third of all SOC reports (36.7%). This principle concerned the availability of systemic measures to address the impacts of climate change. Mentions discussing issues of landscape governance were in third place (18%). This referred to the inclusion of heritage conservation at wider strategic and operational levels as a means for responding to climate change, particularly acknowledging policies, strategies, and the coordination of different management sectors. Finally, landscape scale was the least-discussed principle (8%) and focused on the upscaling of conservation activities beyond WH property boundaries; the principle was also discussed in relation to socioeconomic context and other environmental or physical aspects pertaining to where properties were located.

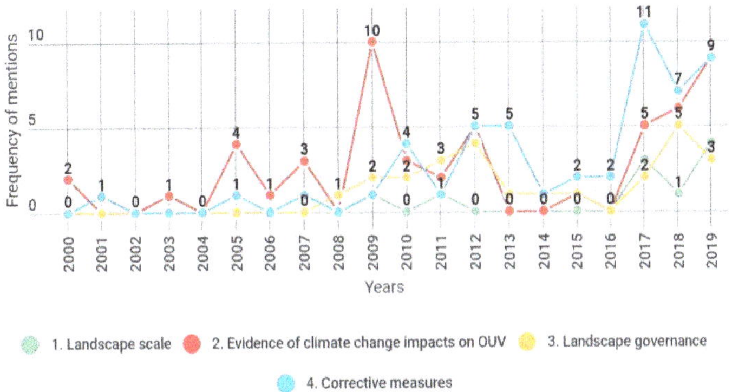

Figure 5. Mentions identified as landscape-based conservation principles (coding dimension 4).

The summative of post-coding dimensions that classify the types of mentions of climate change proved that these became more elaborated over time in taxonomic terms. This can be seen by the increase of concurrences of post-coding items over time. During the period 2009 to 2012, a peak in mentions of natural properties was observed to correlate with more than one item from landscape-based conservation principles and context of mentions. A second peak in correlations was observed from 2017 to 2019. In both periods, natural properties were found to be predominant; in the most recent period, however, cultural properties became more visible (see Figures 4 and 5). Latest discussions were more likely to include several items of post-coding dimensions three and four than earliest mentions. As such, these included the recognition of measures implemented by State Parties together with expressed concerns and provided recommendations. Additionally, each context of mentions could be linked to one or more landscape-based conservation principles. Yet, this does not necessarily mean that discussions on climate change are becoming more comprehensive regarding management practices. This is further explained in the next section. A general summary of findings is shown in Appendix A.

4.3. Relationships between Post-Coding Dimensions

In this section, we present the results of the relationship analysis between post-coding dimensions in hierarchical order. Here, post-coding dimension four, landscape-based principles, as well as related items aligned under items of post-coding dimension three, contexts of mentions.

Figure 6 indicates that the most frequent correlations between post-coding dimensions linked concerns (post-coding item 3.1) with adaptation measures (post-coding item 4.4). Examples of these mentions include common petitioning from the WHCOM to State Parties to reduce the impacts of previously identified threats (e.g., pollution, natural phenomena, natural resources depletion) that had been observed as exacerbated by climate change, and that may potentially increase a property's vulnerability. This situation suggests a lack of management and conservation strategies for reducing the vulnerability of WH properties to climate change, particularly to environmental risks. Following on closely, post-coding item 3.1, expression of concerns, correlated to post-coding item 4.2, evidence of climate change impacts. Considering that WH properties are requested to implement their own monitoring and documentation processes, related mentions highlighted a lack of adequate methods for documentation, data collection, and for assessing the impacts of climate change on properties' OUV as a means for informing and supporting the prioritisation of actions. It was observed that these mentions often lack a definition of the type of threat and how it is related to climate change; for example, an increase in the frequency and/or intensity of natural phenomena, or the transformation of environmental settings as stated in projections for particular local contexts, etc. Finally, concerns (item 3.1) showed the least correlation to post-coding items 4.3 and 4.1, landscape scale and landscape governance, respectively. These groups of mentions focused on understanding properties' components and conservation actions within a larger physical scope and included the involvement of a broader range of actors; for example, the consideration of WH properties at higher policy and management levels that targeted climate change. Examples of mentions found correlating dimension three, one, and dimension four per WH category are shown in the Table 2 item.

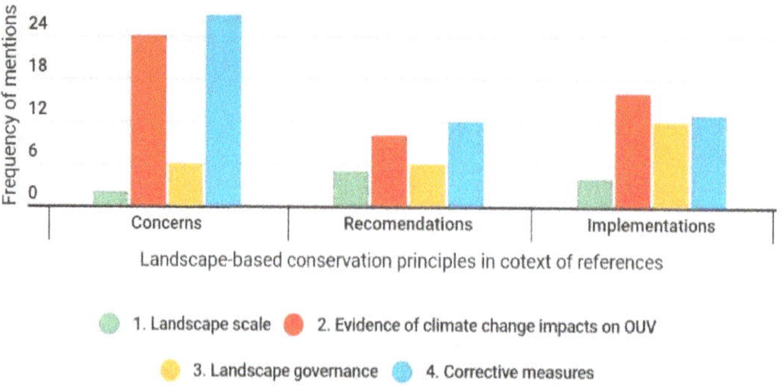

Figure 6. Illustrates the summative assessment of post-coding dimensions three and four.

The second group of correlations between post-coding dimensions, as shown in Figure 6, found an increasing recognition of implementation of measures primarily responding to the need for gathering evidence of the impacts of climate change on WH properties (post-coding items 3.3 and 4.2). More comprehensive monitoring programs and assessment studies are mentioned as requirements within the framework of studies and conservation programs that are primarily undertaken by an entity external to the national party and local management officials. Following on closely was the frequency of mentions for subgroups involving correlations between implementations of corrective measures (post-coding item 4.4) and landscape governance (post-coding item 4.3). These two categories are related because rehabilitation and maintenance activities are often strengthened at higher strategic and management levels; for example, for natural properties, such actions aim at maintaining OUV related to biodiversity. In cultural heritage, actions may include developing new protective infrastructures that can facilitate WH adaptation to climate change. These include considerations for heritage conservation in the form of strategic management and development plans applied at a broader territorial level in which heritage is embedded. The least frequent mention in this group (see Figure 6) correlated implementations with post-coding item 4.4, landscape scale. Related actions included the extension of WH properties' boundaries and the implementation of transboundary studies (e.g., the Great Himalayan National Park Conservation Area and the overall approach to the waterfront in the Historic Inner City of Paramaribo). Examples of this group are shown in Table 3.

The third group of correlations (Figure 6) corresponded to recommendations (post-coding item 3.2) made by the WHC to State Parties regarding climate change management and measures (post-coding item 4.4), followed by regular maintenance and adaptive management measures supported by enhanced monitoring tools and the involvement of wider governance levels (post-coding items 4.2 and 4.3). A final correlation represented acting in collaboration with other sectors for reducing threats and pressures that increased properties' vulnerability to climate change (post-coding item 4.1). Table 4 shows examples of this group of mentions.

Table 2. Summary of main findings per WH category.

Mentions of Climate Change as Contexts of Mentions (Dimensions One and Three)	Dim. 2: SOC Reports' General Information			Dim. 4: Landscape Conservation Principles
Item one: keywords in texts	Item 2	3	4	Item 1–4
"[T]he current lack of effective site management for most components, as well as identified threats. At present, [a] level of protection and management is only available for the three large forts. The other forts are mostly abandoned, with accelerated degradation due to rising salts and humidity, lack of effective management and the possibility of climate change impacts."	2019 Forts and castles, Volta, Greater Accra, Central and Western Regions	C	Ghana, AFR	(1) Landscape scale (2) Evidence of climate change impacts (4) Adaptation measures
"The Old City of Shibam, [constructed] of mud and located in a flood prone area, remains at severe risk of major damage, unless necessary preventive measures are taken. The effects of climate change are exacerbating this problem. Therefore, the proposed 'Shibam Oases Development Project' is essential for long-term conservation. This project, which is part of a sustainable food security program, involves preparation of a comprehensive developmental plan for the conservation and use of Shibam oases, which are considered as the buffer zone of the property."	2018 Old Walled City of Shibam	C	Yemen ARA	(3) Landscape governance (4) Adaptation measures
"According to the report, the site is subject to changes due to socio-economic phenomena (schooling, infrastructure development), human activities and environmental degradation (climate change, demographic pressure). Some intangible cultural practices are mutating due to contact with other imported values (religions, cultural tourism)."	2013 Cliffs of Bandiagara (Land of the Dogons)	C/N	Mali AFR	(4) Adaptation measures
"[The] future integrity of the property [faces high] risk, taking into account the possible prospect of offshore oil exploitation … uncertainty about the impact of invasive species, the already existing threats for which progress on the corrective measures is unclear and the globally increasing effects of climate change to coral reef systems."	2012 Belize Barrier Reef Reserve System	N	Belize LAC	(4) Adaptation measures

Table 3. Example of analysis implementations (item 3.2) and landscape conservation principles (dimension four).

Mentions of Climate Change as Implementations (Dimensions One and Three)	Dim. 2: SOC Reports' General Information			Dim. 4: Landscape Conservation Principles
Item one: keywords in texts	Items 2	3	4	Items one to four
"An India-Bangladesh Joint Working Group on Conservation of the Sundarbans (JWG) has been created, which has led to an agreement to conduct joint studies on the indicators to monitor the health of the Sundarbans ecosystem and the impacts of climate change and to protect the Bengal Tiger."	2019 Sundarbans	N	Bangladesh APA	(1) Landscape scale (2) Evidence of climate change impacts (4) Corrective measures
"Other issues are reported, such as the presence of clandestine interventions in historic buildings, the potential impact of wholesale trade and street markets and the need to improve risk preparedness in the property in order to face El Niño and climate change threats. All issues were taken into account in the elaboration of the new Master Plan and are expected to be addressed with its implementation."	2019 Historic Centre of Lima	C	Peru LAC	(2) Evidence of climate change impacts (3) Landscape governance (4) Corrective measures
"At Eridu, a landscape-based survey of the hinterland of the city is being planned to research the impact of climate change on the marshes, as is a survey of the city and a Conservation Plan; funding is being sought for a five-year research, excavation and enhancement project."	2018 The Ahwar of Southern Iraq: Refuge of Biodiversity and the Relict Landscape of the Mesopotamian Cities	C/N	Iraq ARA	(2) Evidence of climate change impacts (3) Landscape governance

Table 4. Example of analysis recommendations (item 3.3) and landscape conservation principles (dimension four).

Mentions to Climate Change as Recommendations (Dimensions One and Three)	Dim. 2: SOC Reports' General Information			Dim. 4: Landscape Conservation Principles
Item one: keywords in texts	Item 2	3	4	Items one to four
"Several colonies continue to be observed outside the property and given their susceptibility to other factors, including climate change, it is recommended that the Committee encourage the State Party to develop a proposal for an extension of the property in order to ensure that the majority of the areas occupied by overwintering colonies are properly protected and to increase the potential of the property to adapt to changing climatic conditions and associated changes in the distribution of overwintering colonies."	2019 Monarch Butterfly Biosphere Reserve	N	Mexico LAC	(1) Landscape scale (2) Evidence of climate change impacts (4) Corrective measures
"With regard primarily to the issue of tourism pressure and the negative impacts of climate change, the property remains subject to the cumulative impact of ascertained and potential threats. Sufficient improvement in the state of conservation and further progress with mitigation are therefore [still needed] in order to maintain the authenticity and integrity of the property and to protect its OUV to a level that will prevent the property [from being] considered for inscription on the List of World Heritage [Sites] in Danger."	2019 Venice and its lagoon	C	Italy EUR	(2) Evidence of climate change impacts (4) Corrective measures
"The mission noted that the size of the area and the diversity of its ecosystems contribute to its adaptive capacity to climate change. The property would benefit from an active programme for monitoring the impacts of climate change, including … carrying out a vulnerability assessment for both natural and cultural (archaeological) resources and to prepare an adaptation strategy on that basis. This could be integrated within the recommended strategy and action plan for reducing risks to the World Heritage property."	2007 Tasmanian wilderness	C/N	Australia APA	(1) Landscape scale (2) Evidence of climate change impacts (3) Landscape governance

5. Discussion and Conclusions

This study provided an overview of how climate change is discussed within the WH monitoring system (SOC reports). The reliability of this monitoring tool is often challenged by politicised clashes between State Parties, the WH Centre, and advisory bodies. Despite such limitations, this research revealed that progress has been made in terms of recognising climate change as a threat to WH properties. However, addressing this issue has not yet been sufficiently integrated within the monitoring system. Only 64 (6%) of all WH properties mentioned climate change in SOC reports. Among WH categories, natural heritage leads the discussion on climate change. This is explained by the fact that their OUV is based on environmental characteristics that are directly affected by climate change [77,78]. Although cultural heritage properties have gained visibility in recent years, the number of both cultural and mixed properties are unlikely to reflect the reality of current climate challenges worldwide. This is particularly true when considering the increasing number of literature on climate risks and vulnerability assessments [18].

In this study, we identified four principles of landscape-based approaches used as a response to climate change challenges for the conservation of natural, cultural and mixed properties mentioned in SOC reports. These principles were observed predominantly as concerns expressed by the WHCOM regarding the need for adaptive measures and for gathering evidence that is more consistent regarding the impacts of climatic factors on the OUV of properties. Additionally, monitoring and evidence-based management are increasingly being requested. This suggests that monitoring tools requested by the Operational Guidelines for the Implementation of the WHC (and as part of the process for a property to be nominated with WH status) are likely insufficient for assessing the complexity of climate change in relation to WH conservation.

Conservation using a landscape approach provides considerable room for action regarding collaborative strategies among governance and managerial sectors at the local level. Our results showed that both the upscaling of conservation activities in its wider context, as well as partnerships among different sectoral and governance sectors, can benefit from additional exploration in the context of climate change solutions. Our analysis also identified that a clarification is required regarding two aspects of what is being monitored and reported as 'climate change' in SOC reports. The first aspect refers to the identification of threats resulting from climate change, e.g., the increase in frequency and/or intensity of given natural phenomena. The second aspect refers to the transformation of environmental settings as part of the earth's natural processes related to climate change. Differentiation in the reporting and monitoring of such situational aspects raises two very different challenges for the conservation discipline. Increasing natural phenomena and their related risks challenge local managerial capacities and calls for adaptation strategies and risk preparedness. Furthermore, the transformation of the planet's physical environment can (and should) lead to deeper questioning of the paradigm behind the conservation discipline. According to Solli et al. [79], 'The more fundamental (almost existential) question is what will climate change do to the concept of heritage and our way of expressing scientific narratives about the past?'

We conclude that landscape approaches require additional study not only for the management of WH properties but also for the expansion of heritage conceptualisations. Climate change makes evident the need for deeper connections between cultural and natural values. Thus, heritage understandings should be expanded to include local, regional and/or State significance, and into the relationships between nature and culture. In this sense, developing a common understanding of what constitutes anthropogenic climate change and its associated impacts, including how these can be reduced through feasible climate adaptation actions for both natural and cultural properties, may serve as a crucial first step in improving the monitoring of WH properties. As highlighted in existing research [49,80], a new or revised operationalisation of the limits of acceptable change, without losing heritage values, is crucial for the climate adaptation of WH properties. However, this responsibility should not fall solely on the UNESCO WH system. Rather, it should be undertaken primarily by State Parties as part of their own agendas on mitigation and adaptation to climate change. For policymakers, the exploration and implementation of landscape-based management approaches can advance the adaptation and strengthening of WH multi-scale governance systems. This can foster the operationalisation of local sustainable goals that can integrate heritage values, while also enhancing good governance practices.

Furthermore, adapting comparative methods such as those used by qualitative datasets for policy analysis [81] can help to uncover useful pathways for UNESCO agendas. For example, using the concept of saliency can retrospectively reconstruct policy agendas, compare agendas over time, enable scholars to reconstruct relative attention to all issues on an agenda, and benefit the monitoring system with a more transparent and systematic analysis for all WH properties. This will allow for embedding WH contexts within larger environmental and political processes of climate adaptation and mitigation.

In the meantime, UNESCO is presently updating its climate change policy for WH properties (see https://whc.unesco.org/en/climatechange/). The updated policy is expected to promote preventive and proactive measures for implementation by State Parties in order to strengthen the resilience of cultural and natural WH properties. The landscape principles discussed in this paper can serve as a basis for addressing the gaps in practice and knowledge related to the climate change management of WH properties and address related impacts through actions for adaptation and mitigation.

Author Contributions: Conceptualization, P.G.; revisions and refinement of methodology, S.F. and M.I.; validation, S.F. and M.I.; formal analysis, P.G.; resources, all authors; data curation, P.G.; writing—original draft preparation, P.G.; writing—review and editing, all authors; visualization, P.G.; project administration, P.G.; funding acquisition, NIKU. All authors have read and agreed to the published version of the manuscript.

Funding: This work was supported by the Research Council of Norway [160010/F40].

Conflicts of Interest: The authors declare no conflict of interest. The funders had no role in the design of the study; in the collection, analyses, or interpretation of data; in the writing of the manuscript, or in the decision to publish the results.

Appendix A

Table A1. Summary of the total number of mentions per WH category (does not include the total number of WH properties).

Coding Dimension	Coding Items	Total Mentions	WH Categories (Dimension Two, Item Three)		
			Natural	Cultural	Mixed
1. Freq. of mentions	1. No. of times climate change is mentioned (mentions)	103	83	14	6
2. SOC reports' general information	APA	36	32	3	1
	EUR	28	24	3	1
	AFR	24	17	4	3
	LAC	12	9	4	0
	ARA	3	1	1	1
3. Context of mentions	1. Concerns (C)	49	38	8	3
	2. Recommendations (R)	22	20	1	1
	3. Implementations (I)	32	25	5	2
4. Landscape-based conservation principles	1. Landscape scale	11	8	2	1
	2. Evidence of climate change impacts on OUV	52	44	5	3
	3. Landscape governance	25	16	7	2
	4. Adaptation measures	51	37	12	2

References

1. IPCC. *Fifth Assessment Report: Climate Change*; IPCC: Geneva, Switzerland, 2014.
2. Heathcote, J.; Fluck, H.; Wiggins, M. Predicting and Adapting to Climate Change: Challenges for the Historic Environment. *Hist. Environ. Policy Pract.* **2017**, *8*, 89–100. [CrossRef]
3. ICOMOS. *The Future of Our Pasts: Engaging Cultural Heritage in Climate Action*; ICOMOS: Paris, France, 2019.

4. UNESCO. Climate Change and World Heritage: Report on predicting and managing the impacts of climate change on World Heritage and strategy to assist State Parties to implement appropriate management responses. *World Herit. Rep.* **2007**, *22*, 1–55.

5. UNESCO. *Policy Document on the Impacts of Climate Change on World Heritage Properties*; UNESCO: Paris, France, 2008.

6. Harkin, K.; Davies, D.; Hyslop, M.; Fluck, E.; Wiggins, H.; Merritt, M.; Barker, O.; Deery, L.; McNeary, M.; Westley, R. Impacts of climate change on cultural heritage. *MCCIP Sci. Rev.* **2020**, *16*, 24–39.

7. Fatorić, S.; Seekamp, E. Evaluating a decision analytic approach to climate change adaptation of cultural resources along the Atlantic Coast of the United States. *Land Use Policy* **2017**, *68*, 254–263. [CrossRef]

8. Heron, S.F.; Eakin, C.M.; Douvere, F.; Anderson, K.; Day, J.C.; Geiger, E.; Hoegh-Guldberg, O.; van Hooidonk, R.; Hughes, T.; Obura, D.O. *Impacts of Climate Change on World Heritage Coral Reefs: A First Global Scientific Assessment*; UNESCO: Paris, France, 2017; pp. 1–14.

9. Sesana, E.; Gagnon, A.S.; Bonazza, A.; Hughes, J.J. An integrated approach for assessing the vulnerability of World Heritage Sites to climate change impacts. *J. Cult. Herit.* **2020**, *41*, 211–224. [CrossRef]

10. Perry, J. Climate Change Adaptation in Natural World Heritage Sites: A Triage Approach. *Climate* **2019**, *7*, 105. [CrossRef]

11. Hall, C.M.; Baird, T.; James, M.; Ram, Y. Climate change and cultural heritage: Conservation and heritage tourism in the anthropocene. *J. Herit. Tour.* **2016**, *11*, 10–24. [CrossRef]

12. UNESCO. *Operational Guidelines for the Implementation of the World Heritage Convention*; UNESCO: Paris, France, 2015.

13. Pendlebury, J.; Short, M.; While, A. Urban World Heritage Sites and the problem of authenticity. *Cities* **2009**, *26*, 349–358. [CrossRef]

14. Plieninger, T.; Kizos, T.; Bieling, C.; Le Dû-Blayo, L.; Budniok, M.A.; Bürgi, M.; Kuemmerle, T.; Crumley, C.L.; Girod, G.; Howard, P. Exploring ecosystem-change and society through a landscape lens: Recent progress in european landscape research. *Ecol. Soc.* **2015**, *20*, 1–10. [CrossRef]

15. Zamarbide Urdaniz, A.V. Regional Heritage Dimensions vs. Management Boundaries. *Int. Rev. Spat. Plan. Sustain. Dev.* **2018**, *6*, 64–81.

16. Badman, T.; Kormos, C.F.; Jaeger, T.; Bertzky, B.; van Merm, R.; Osipova, E.; Larsen, P.B. *World Heritage, Wilderness and Large Landscapes and Seascapes*; IUCN International Union for Conservation of Nature: Gland, Suisse, 2017.

17. ICOMOS. *Guidance on Heritage Impact Assessments for Cultural World Heritage Properties*; ICOMOS: Paris, France, 2011.

18. Fatorić, S.; Seekamp, E. Are cultural heritage and resources threatened by climate change? A systematic literature review. *Clim. Chang.* **2017**, *142*, 227–254. [CrossRef]

19. Brown, N.E.; Liuzza, C.; Meskell, L. The Politics of Peril: UNESCO's List of World Heritage in Danger. *J. Field Archaeol.* **2019**, *44*, 287–303. [CrossRef]

20. Meskell, L. States of conservation: Protection, politics, and pacting within UNESCO's world heritage committee. *Anthropol. Q.* **2014**, *87*, 217–243. [CrossRef]

21. UNESCO. *UNESCO Recommendation on the Historic Urban Landscape*; UNESCO: Paris, France, 2011.

22. UNESCO. *Policy Document for the Integration of a Sustainable Development Perspective into the Processes of the World Heritage Convention*; UNESCO: Paris, France, 2015.

23. Osipova, E.; Wilson, L.; Blaney, R.; Shi, Y.; Fancourt, M.; Strubel, M.; Verschuuren, B. *The Benefits of Natural World Heritage: Identifying and Assessing Ecosystem Services and Benefits Provided by the World's Most Iconic Natural Places*; IUCN: Gland, Switzerland, 2014.

24. Adger, W.N.; Barnett, J.; Brown, K.; Marshall, N.; O'Brien, K. Cultural dimensions of climate change impacts and adaptation. *Nat. Clim. Chang.* **2013**, *3*, 112–117. [CrossRef]

25. Reed, J.; Van Vianen, J.; Deakin, E.L.; Barlow, J.; Sunderland, T. Integrated landscape approaches to managing social and environmental issues in the tropics: Learning from the past to guide the future. *Glob. Chang. Biol.* **2016**, *22*, 2540–2554. [CrossRef]

26. Bandarin, F.; van Oers, R. *The Historic Urban Landscape: Managing Heritage in an Urban Century*; Wiley-Blackwell: Oxford, UK, 2012; Volume 3.

27. Meskell, L.; Liuzza, C.; Bertacchini, E.; Saccone, D. Multilateralism and UNESCO World Heritage: Decision-making, States Parties and political processes. *Int. J. Herit. Stud.* **2015**, *21*, 423–440. [CrossRef]

28. Turner, M.; Roders, A.P.; Patry, M. Revealing the Level of Tension Between Cultural Heritage and Development in World Heritage Cities. *Probl. Sustain. Dev.* **2012**, *7*, 23–31.

29. Guzman, P.; Roders, A.P.; Colenbrander, B. Impacts of Common Urban Development Factors on Cultural Conservation in World Heritage Cities: An Indicators-Based Analysis. *Sustainability* **2018**, *10*, 853. [CrossRef]

30. Ishizawa, M. Cultural Landscapes Link to Nature: Learning from Satoyama and Satoumi. *Built Herit.* **2018**, *4*, 7–19.

31. Linnell, J.D.C.; Kaczensky, P.; Wotschikowsky, U.; Lescureux, N.; Boitani, L. Framing the relationship between people and nature in the context of European conservation. *Conserv. Biol.* **2015**, *29*, 978–985. [CrossRef]

32. Descola, P.; Lloyd, J.; Sahlins, M. *Beyond Nature and Culture*; University of Chicago Press: Chicago, IL, USA, 2013.

33. Descola, P.; Pálsson, G. *Nature and Society: Anthropological Perspectives*; Taylor & Francis: Abingdon-on-Thames, UK, 1996.

34. Fatorić, S.; Seekamp, E. Securing the Future of Cultural Heritage by Identifying Barriers to and Strategizing Solutions for Preservation under Changing Climate Conditions. *Sustainability* **2017**, *9*, 2143. [CrossRef]

35. Simensen, T.; Halvorsen, R.; Erikstad, L. Methods for landscape characterisation and mapping: A systematic review. *Land Use Policy* **2018**, *75*, 557–569. [CrossRef]

36. Smith, L. *Uses of Heritage*; Routledge: Abingdon-on-Thames, UK, 2006.

37. Pfund, J.L. Landscape-scale research for conservation and development in the tropics: Fighting persisting challenges. *Curr. Opin. Environ. Sustain.* **2010**, *2*, 117–126. [CrossRef]

38. Brunetta, G.; Voghera, A. Evaluating landscape for shared values: Tools, principles, and methods. *Landsc. Res.* **2008**, *33*, 71–87. [CrossRef]

39. Wu, J. Landscape sustainability science: Ecosystem services and human well-being in changing landscapes. *Landsc. Ecol.* **2013**, *28*, 999–1023. [CrossRef]

40. Gustafson, E.J. Quantifying landscape spatial pattern: What is the state of the art? *Ecosystems* **1998**, *1*, 143–156. [CrossRef]

41. Tscharntke, T.; Tylianakis, J.M.; Rand, T.A.; Didham, R.K.; Fahrig, L.; Batáry, P.; Bengtaaon, J.; Clough, Y.; Crist, T.O.; Ewers, R.M. Landscape moderation of biodiversity patterns and processes—Eight hypotheses. *Biol. Rev.* **2012**, *87*, 661–685. [CrossRef]

42. Bandarin, F.; van Oers, R. (Eds.) *Reconnecting the City: The Historic Urban Landscape Approach and the Future of Urban Heritage*; Wiley-Blackwell: Oxford, UK, 2014.

43. Roders, A.P.; Bandarin, F. (Eds.) *Reshaping Urban Conservation: The Historic Urban Landscape Approach in Action*; Springer: Singapore, 2019.

44. Veldpaus, L. *Historic Urban Landscapes, Framing the Integration of Urban and Heritage Planning in Multilevel Governance*; Eindhoven University of Technology: Eindhoven, The Netherlands, 2015.

45. Van Oers, R.; Roders, A.P. Aligning agendas for sustainable development in the post 2015 world. *J. Cult. Herit. Manag. Sustain. Dev.* **2014**, *4*, 122–132. [CrossRef]

46. Sayer, J.; Sunderland, T.; Ghazoul, J.; Pfund, J.L.; Sheil, D.; Meijaard, E.; Van Oosten, C. Ten principles for a landscape approach to reconciling agriculture, conservation, and other competing land uses. *Proc. Natl. Acad. Sci. USA* **2013**, *110*, 8349–8356. [CrossRef]

47. Aplin, G. World heritage cultural landscapes. *Int. J. Herit. Stud.* **2007**, *13*, 427–446. [CrossRef]

48. Olwig, K.R.; Dalglish, C.; Fairclough, G.; Herring, P. Introduction to a special issue: The future of landscape characterisation, and the future character of landscape—Between space, time, history, place and nature. *Landsc. Res.* **2016**, *41*, 169–174. [CrossRef]

49. Veldpaus, L.; Roders, A.P.; Colenbrander, B.J.F. Urban Heritage: Putting the Past into the Future. *Hist. Environ.* **2013**, *4*, 18–33. [CrossRef]

50. Verburg, P.H.; van Asselen, S.; van der Zanden, E.H.; Stehfest, E. The representation of landscapes in global scale assessments of environmental change. *Landsc. Ecol.* **2013**, *28*, 1067–1080. [CrossRef]

51. Tress, B.; Tress, G.; Décamps, H.; D'Hauteserre, A.M. Bridging human and natural sciences in landscape research. *Landsc. Urban Plan.* **2001**, *57*, 137–141. [CrossRef]

52. Taylor, K.; Lennon, J. Cultural landscapes: A bridge between culture and nature? *Int. J. Herit. Stud.* **2011**, *17*, 537–554. [CrossRef]

53. Aylett, A. Institutionalizing the urban governance of climate change adaptation: Results of an international survey. *Urban Clim.* **2015**, *14*, 4–16. [CrossRef]

54. Phillips, H. The capacity to adapt to climate change at heritage sites-The development of a conceptual framework. *Environ. Sci. Policy* **2015**, *47*, 118–125. [CrossRef]

55. Boyle, M.; Kay, J.; Pond, B. Monitoring in support of policy: An adaptive ecosystem approach. *Encycl. Glob. Environ. Chang.* **2001**, *4*, 116–137.

56. Hou, W.; Walz, U. Enhanced analysis of landscape structure: Inclusion of transition zones and small-scale landscape elements. *Ecol. Indic.* **2013**, *31*, 15–24. [CrossRef]

57. Volpiano, M. Indicators for the Assessment of Historic Landscape Features. In *Landscape Indicators*; Cassatella, C., Peano, A., Eds.; Springer: Dordrecht, The Netherlands, 2011; pp. 77–104.

58. Tengberg, A.; Fredholm, S.; Eliasson, I.; Knez, I.; Saltzman, K.; Wetterberg, O. Cultural ecosystem services provided by landscapes: Assessment of heritage values and identity. *Ecosyst. Serv.* **2012**, *2*, 14–26. [CrossRef]

59. Cassatella, C.; Peano, A. Indicators for the Assessment of Historic Landscape Features. In *Landscape Indicators*; Cassatella, C., Peano, A., Eds.; Springer: Dordrecht, The Netherlands, 2011; Volume 53, pp. 1–30.

60. Stem, C.; Margoluis, R.; Salafsky, N.; Brown, M. Monitoring and evaluation in conservation: A review of trends and approaches. *Conserv. Biol.* **2005**, *19*, 295–309. [CrossRef]

61. Bond, A.; Langstaff, L.; Baxter, R.; Kofoed, H.-G.W.J.; Lisitzin, K.; Lundström, S. Dealing with the cultural heritage aspect of environmental impact assessment in Europe. *Impact Assess. Proj. Apprais.* **2004**, *22*, 37–45. [CrossRef]

62. Rössler, M. World Heritage cultural landscapes: A UNESCO flagship programme 1992–2006. *Landsc. Res.* **2006**, *31*, 333–353. [CrossRef]

63. Ahern, J. Urban landscape sustainability and resilience: The promise and challenges of integrating ecology with urban planning and design. *Landsc. Ecol.* **2013**, *28*, 1203–1212. [CrossRef]

64. Sunderland, T.C.H.; Ehringhaus, C.; Campbell, B.M. Conservation and development in tropical forest landscapes: A time to face the trade-offs? *Environ. Conserv.* **2007**, *34*, 276–279. [CrossRef]

65. Fatorić, S.; Seekamp, E. A measurement framework to increase transparency in historic preservation decision-making under changing climate conditions. *J. Cult. Herit.* **2018**, *30*, 168–179. [CrossRef]

66. Reed, J.; van Vianen, J.; Barlow, J.; Sunderland, T. Have integrated landscape approaches reconciled societal and environmental issues in the tropics? *Land Use Policy* **2017**, *63*, 481–492. [CrossRef]

67. Xiao, X.; Seekamp, E.; van der Burg, M.P.; Eaton, M.; Fatorić, S.; McCreary, A. Optimizing historic preservation under climate change: Decision support for cultural resource adaptation planning in national parks. *Land Use Policy* **2019**, *83*, 379–389. [CrossRef]

68. Haugen, A.; Bertolin, C.; Leijonhufvud, G.; Olstad, T.; Broström, T. A Methodology for Long-Term Monitoring of Climate Change Impacts on Historic Buildings. *Geosciences* **2018**, *8*, 370. [CrossRef]

69. Heilen, M.; Altschul, J.H.; Lüth, F. Modelling Resource Values and Climate Change Impacts to Set Preservation and Research Priorities. *Conserv. Manag. Archaeol. Sites* **2018**, *20*, 261–284. [CrossRef]

70. Fouseki, K.; Nicolau, M. Urban Heritage Dynamics in 'Heritage-Led Regeneration': Towards a Sustainable Lifestyles Approach. *Hist. Environ. Policy Pract.* **2018**, *9*, 229–248. [CrossRef]

71. Guzmán, P.C.; Roders, A.R.P.; Colenbrander, B.J.F. Measuring links between cultural heritage management and sustainable urban development: An overview of global monitoring tools. *Cities* **2017**, *60*, 192–201. [CrossRef]

72. Harvey, D.; Perry, J. (Eds.) *The Future of Heritage as Climates Change*; Routledge: London, UK, 2015.

73. Garcia, B.M. Resilient cultural heritage: From global to national levels—The case of Bhutan. *Disaster Prev. Manag. An. Int. J.* **2019**, *29*, 36–46. [CrossRef]

74. O'Brien, G.; O'Keefe, P.; Jayawickrama, J.; Jigyasu, R. Developing a model for building resilience to climate risks for cultural heritage. *J. Cult. Herit. Manag. Sustain. Dev.* **2015**, *5*, 99–114. [CrossRef]

75. Daly, C. The design of a legacy indicator tool for measuring climate change related impacts on built heritage. *Herit. Sci.* **2016**, *4*, 19. [CrossRef]

76. Young, C. Understanding management in a world heritage context: Key current issues in Europe. *Hist. Environ. Policy Pract.* **2016**, *7*, 189–201. [CrossRef]

77. Perry, J. World Heritage hot spots: A global model identifies the 16 natural heritage properties on the World Heritage List most at risk from climate change. *Int. J. Herit. Stud.* **2011**, *17*, 426–441. [CrossRef]

78. Perry, J. Climate change adaptation in the world's best places: A wicked problem in need of immediate attention. *Landsc. Urban Plan.* **2015**, *133*, 1–11. [CrossRef]

79. Solli, B.; Burström, M.; Domanska, E.; Edgeworth, M.; Gonz'alez-Ruibal, A.; Holtorf, C.; Lucas, G.; Oestigaard, T.; Smith, L. Some Reflections on Heritage and Archaeology in the Anthropocene. *Nor. Archaeol. Rev.* **2011**, *44*, 40–88. [CrossRef]

80. Jokilehto, J. Considerations on authenticity and integrity in World Heritage context. *City Time* **2006**, *2*, 1–16.

81. Alexandrova, P.; Carammia, M.; Princen, S.; Timmermans, A. Measuring the European Council agenda: Introducing a new approach and dataset. *Eur. Union Polit.* **2013**, *15*, 152–167. [CrossRef]

Article

An Ecolabel for the World Heritage Brand? Developing a Climate Communication Recognition Scheme for Heritage Sites

Kathryn Lafrenz Samuels [1,*] **and Ellen J. Platts** [1]

Department of Anthropology, University of Maryland, College Park, MD 20742, USA;
ejplatts@terpmail.umd.edu
* Correspondence: lafrenzs@umd.edu

Received: 30 January 2020; Accepted: 3 March 2020; Published: 5 March 2020

Abstract: This study develops a climate communication recognition scheme (CCRS) for United Nations Educational, Scientific and Cultural Organization (UNESCO) World Heritage Sites (WHS), in order to explore the communicative power of heritage to mobilize stakeholders around climate change. We present this scheme with the aim to influence site management and tourist decision-making by increasing climate awareness at heritage sites and among visitors and encouraging the incorporation of carbon management into heritage site management. Given the deficits and dysfunction in international governance for climate mitigation and inspired by transnational environmental governance tools such as ecolabels and environmental product information schemes, we offer "climate communication recognition schemes" as a corollary tool for transnational climate governance and communication. We assess and develop four dimensions for the CCRS, featuring 50 WHS: carbon footprint analysis, narrative potential, sustainability practices, and the impacts of climate change on heritage resources. In our development of a CCRS, this study builds on the "branding" value and recognition of UNESCO World Heritage, set against the backdrop of increasing tourism—including the projected doubling of international air travel in the next 15–20 years—and the implications of this growth for climate change. The CCRS, titled *Climate Footprints of Heritage Tourism*, is available online as an ArcGIS StoryMap.

Keywords: climate change; climate communication; mitigation; adaptation; World Heritage; heritage tourism; carbon footprint; carbon management; ecolabel; environmental product information scheme; transnational governance

1. Introduction

Heritage tourism represents an important nexus in which cultural heritage resources and anthropogenic climate change intersect. This is especially the case for efforts pursuing climate mitigation and climate communication. Tourism is a major contributor to greenhouse gas emissions, and the tourism industry will require broad changes to operations in order to mitigate climate change [1–4]. In 2005, the tourism industry (including transport, accommodation, and activities) was estimated to contribute 5% of global anthropogenic carbon dioxide emissions (with aviation transport accounting for 40% of this amount, car transport for 32%, and accommodation for 21%) [5–7]. By 2013, tourism accounted for 8% of global greenhouse gas emissions, and previous reports may have underestimated the contribution [8]. Moreover, tourism is only expected to grow, with carbon dioxide emissions predicted to increase 135% between 2005 and 2035 [5–7] and international air travel projected to double in the next 15–20 years [9]. Scholars in tourism studies have turned a spotlight on global climate change [2,3,10,11], including increasingly sophisticated accounts of the carbon systems in tourism [12,13] and "carbon management" to mitigate the impacts of climate change from tourism [1]. The impetus of our study dovetails with current research in tourism studies on climate change, but cultural heritage has rarely been the focus of such studies.

Heritage sites and resources feed the trends in the diversification and geographic expansion of tourism. Indeed, the case of heritage tourism is especially urgent, since according to some estimates it accounts for approximately half of all tourism [5,14], and it contributes to the destruction of the resources on which it is based, both presently and, considering projected climate impacts, in the future. Moreover, taking into account that heritage tourism is one of the principle means for pursuing economic development through heritage resources, the detrimental impacts of heritage tourism on heritage resources in light of climate change thus raises the question of whether the use of heritage tourism for sustainable development is viable.

At the same time, World Heritage has also been picked up by climate scientists and activists as a flashpoint for taking action on global climate change, given the impacts that climate change will have on well-known heritage sites beloved by many [15,16]. The Great Barrier Reef has attracted particular attention, as its status as a World Heritage Site (subject to international treaty law) has been leveraged to bring the Australian Government to account for its tepid response to climate change mitigation [17,18]. An explicit focus on cultural heritage in climate action and climate communication can raise in sharp relief the anthropogenic and long-term character of global climate change. Yet, as the link between global climate change and loss of WHS is made stronger in the public imagination, trends in what is called 'last chance tourism' [19–22] suggest that WHS impacted by climate change will attract increasing numbers of visitors, to witness the WHS 'before it is gone' or irreparably changed. 'Last chance tourism' often has the paradoxical effect of hastening the deterioration of the site in question, for example from increasing pollution and greenhouse gas emissions from Antarctic cruise tourism [23] or the carbon costs associated with polar bear tourism in Churchill, Canada [24].

National governments and local development initiatives often pursue inscription of sites on the UNESCO World Heritage List because of its performance in attracting international tourism (the most "costly" of tourism carbon footprints) [25–29]. Indeed, the UNESCO World Heritage List has exploded to over 1100 properties and counting, overwhelming the List's original conservation mandate as national governments seek to attract foreign tourism revenue and foster economic growth through the World Heritage "brand." The branding function of World Heritage is powerful and increasingly central to motivations for inscription [30–33].

WHS around the globe hold a special resonance for the communities residing within or nearby the sites, the governments that nominate the sites at much expenditure, and the tourists and other visitors who choose a World Heritage Site as a destination worth visiting. As the world's cultural treasures, WHS elicit concern for conservation and sustainability, coupled with a well-known brand recognition. These two facets of World Heritage—conservation and branding—provide an excellent vehicle for drawing public awareness and action to the problem of global climate change. Further, while previous sustainability-oriented work on WHS has focused on "sustainable tourism" [34–36], social sustainability [37], or site sustainability [38], scholarship on the sustainability of WHS vis-à-vis climate change remains incipient, especially with respect to the contributions of World Heritage Site facilities and tourism to the production of greenhouse gases and to incorporating carbon management as part of site management and heritage management more broadly.

The growth of the World Heritage List to over 1100 sites also tracks a sea change in heritage management in global contexts more broadly, from management models built around preservation and conservation to those built around economic development [39,40]. Conservation and development have been two sides of the same coin since at least the 1940s, with the rise of international cooperation over heritage resources. However, development is increasingly the raison d'être of heritage management globally, as state and community interest in using cultural heritage for economic development has grown, and international development circles have adopted cultural heritage as a socially-responsive approach to development attuned to sustainability and human capabilities.

In the following we detail the development of what we call a "climate communication recognition scheme" (CCRS) for UNESCO WHS, as a tool for using heritage sites to communicate the drivers and impacts of anthropogenic climate change. The CCRS, titled *Climate Footprints of Heritage Tourism*, is available online as an ArcGIS StoryMap at http://www.heritageofclimate.com [41]. While considerations of climate change are increasingly being incorporated into heritage management practices, including within WHS nominations, the focus remains on climate-related threats and impacts to heritage resources, rather than the contributions of heritage sites as drivers of anthropogenic climate change—for example from site infrastructure, tourism, and land use. We argue for the value of a more holistic response in heritage management to the challenges of climate change, one that takes into account impacts alongside fostering adaptive capacity, integrating carbon management practices for mitigation, and communicating climate change through the public platforms afforded by heritage resources. Given the relative lack of attention paid to mitigation and climate communication, we focus our efforts in the CCRS especially on these two dimensions, but impacts and adaptation are also represented.

This holistic approach that we advocate is further reflected in our decision to conceive of the category of CCRS. We describe CCRS as an extension or elaboration of ecolabels. "Ecolabels" assign labels to consumer goods (for example, organic and fair trade certification of foods) for the purpose of communicating the sustainability of a product, but they tend to be one-dimensional and quantitatively driven. As discussed below, a CCRS expands on ecolabels to provide a more holistic recognition scheme, and one that is specifically attuned to communicating climate change.

While a CCRS could be developed for any heritage site, in this study we focus on WHS to leverage the brand recognition of World Heritage—as global heritage conservator—into a "product" or platform for developing the CCRS. The value of a CCRS for heritage resources is that it offers another frontline for climate action, as a tool for transnational climate governance. While national governments and the international (intergovernmental) system continue to show, on the whole, weak and ineffectual gains in reducing the drivers of climate change, the possibility of advancing climate action through transnational governance channels shows increasing promise. Through the development of a CCRS as a transnational governance tool for heritage management and heritage tourism, we aim to highlight how heritage resources may be mobilized for responding to the challenges of anthropogenic climate change.

2. Materials and Methods

2.1. Identifying a Subset of WHS for Analysis

Recognizing the scale of the World Heritage List—composed of more than 1100 sites and with additional sites being inscribed every year—we began this study as a pilot project and needed to identify a subset of sites for analysis and development into the CCRS. We undertook a broad survey of WHS, and a subset of sites ($n = 50$) was chosen based on the following considerations: (1) most visited and iconic sites; (2) sites impacted heavily by climate change (currently or projected); (3) sites that tell important stories about anthropogenic climate change; (4) sites that have active sustainability initiatives or programs in progress; and (5) representativeness, for example in geographic distribution and site types.

Most visited and iconic sites (e.g., Angkor, the Historic Ensemble of the Potala Palace, Lhasa, and Machu Picchu) were identified through articles in media outlets such as National Geographic and through surveys conducted by TripAdvisor, a popular source of advice for travelers. These sites were included so as to increase interest in and relevance of the results of the analysis, and because as some of the most visited sites they likely generate a disproportionate share of carbon emissions from heritage tourism compared to other sites on the World Heritage List. Sites most at risk of being impacted by climate change were identified by drawing on existing studies by UNESCO World Heritage and its advisory bodies, as well as heritage experts and non-governmental organizations [18,42–45].

Attention to narrative is recognized as an increasingly vital tool for responding to the challenges of climate change [46,47]. Narratives provide conceptual scaffolding not only for adapting to climate change and mitigating its drivers, but also for mobilizing broader conversations about the moral and ethical foundations of a changing climate, and for individually processing how we face climate change in dignity, grief, sorrow, and hope. We identified sites with the narrative potential to tell stories about climate change through a process of qualitative coding [48] of the site descriptions and site documentation for the WHS [49], and associated literature, and coded for themes related to climate change. Coding allowed us to draw out those sites that offered especially holistic (integrating facets such as impacts, adaptation, mitigation, and communication) and visible cases of the nexus of heritage and climate change, as well as themes that might not be captured in the other categories under consideration. For example, sites associated with the use of fossil fuels (e.g., Cornwall and West Devon Mining Landscape, UK) help illustrate the anthropogenic contributions of carbon-based energy sources to climate change, and sites endangered by conflict influenced by climate change (e.g., Palmyra, Syria) highlight that impacts on heritage resources are not only natural or environmental (e.g., sea level rise, increased pest and mold activity) but social too.

Sites with active sustainability practices were identified through an examination of relevant documentation from UNESCO, academic literature, and through news media coverage [50,51]. Finally, sites were added to ensure that the subset was representative of the World Heritage List, with regard to factors such as type, purpose, and geographical location.

The subset of 50 sites was then organized into four subgroups: (a) sites selected for carbon footprint analysis; (b) sites with narrative potential; (c) sites with active sustainability practices; and (d) sites most impacted by climate change (see Appendix A). Some WHS were included in more than one subgroup.

2.2. Calculating Carbon Footprints of Travel

Ten sites were selected for calculating the carbon footprint of travel, chosen to ensure geographic and site type diversity. This study took a consumer-based approach, which allocates the emissions from tourism to the tourist, in recognition of the consumer-tourist as the main driver behind demand [12,13,52,53]. The consumption-based approach estimates the carbon footprint of at minimum three typical visits (budget, mid-range, and high-end spending) to each of these sites. For some sites, the carbon footprints for additional kinds of trips was calculated (e.g., cruise vs. land-based trips to the Galápagos Islands). For the purposes of this study, a typical trip was considered to be one in which a tourist traveled to a destination only to visit the WHS in question. Broad expenditure categories that make up the activities of tourists at these sites include local transport, accommodation, restaurant services, and recreation and leisure services, including heritage tourism activities. Excluded from the analysis was money spent on retail, such as gifts and souvenirs, and any expenditure related to travel to the point of departure in the tourist's home country. The data collected to build out an itinerary for the trip and calculate expenditures were collected from publicly-available websites used by travelers for the purpose of planning trips. These include WorldHeritageSite.org, the official UNESCO World Heritage website, including the individual webpages associated with each site (for identifying activities), Bookings.com (for collecting accommodation prices), TripAdvisor.com (for collecting price information for dining and additional activity costs, Google Flights (for identifying common flight paths), and the International Civil Aviation Organization (ICAO) Carbon Emissions Calculator (for calculating the carbon footprint of international flights).

The carbon footprints were then calculated using an input–output life cycle assessment (IO LCA) model. This model uses monetary transactions to approximate all of the material and energy inputs of a product, good, or service, and their direct and indirect emissions. A United States economy-based IO LCA model was used to calculate the carbon footprint of tourism to each site, due to the relative ease of gathering monetary expenditure data, the expansiveness of the model in terms of industry sectors, and the model's comprehensiveness in terms of incorporating upstream and indirect emissions in the carbon footprint (United States industry-based IO LCA 2002 Purchaser Price model). Expenditure data were assigned to one of the sectors in the model, and emissions for the total expenditure per sector were calculated and summed to get the total carbon footprint for each type of visit, excluding international air travel (Figure 1).

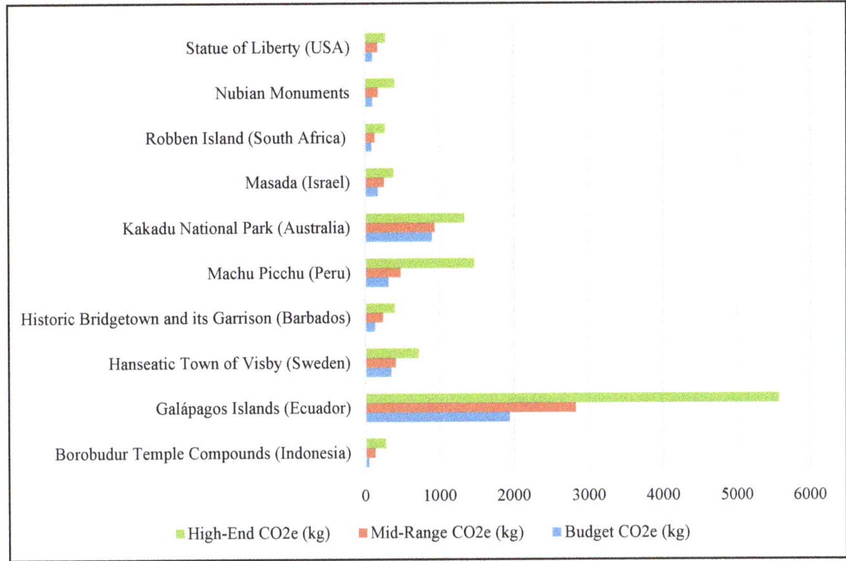

Figure 1. The average carbon dioxide equivalent (CO2e) emissions for three types of trips to selected WHS: High-End, Mid-Range, and Budget. For reference, 100 kilograms of CO2e emissions is equivalent to the carbon sequestered by growing 1.7 tree seedlings for 10 years [54]. For some sites in the CCRS we calculated multiple kinds of trips for each type (for example, multiple mid-range options using different forms of transportation), but in the above graph we have averaged these in the relevant expenditure category (High-End, Mid-Range, or Budget) to simplify the graphical representation.

To improve accuracy, this model was combined with directly calculated emissions data for international flights, using the ICAO Carbon Emissions Calculator [55]. This calculator employs industry averages for the various factors that contribute to the carbon footprint of air travel. It takes into account passenger load factors, passenger-to-cargo ratios, and fuel consumption and calculates an average fuel burn per economy class passenger. The IO LCA results were added to the ICAO international air travel results in order to calculate the total carbon footprint of each type of trip to the WHS, presented in kilograms of CO2e/trip (for an example of one site, depicting travel to Machu Picchu, see Figure 2).

It should be noted that the model used in this study is limited in its ability to accurately calculate emissions based on non-U.S. dollar expenditures, and expenditures were adjusted for inflation from 2002 to 2018. We therefore emphasize that the result is an estimated carbon footprint, used for educational purposes only. Unlike ecolabels, the recognition scheme developed here does not aim to provide rewards or certification to particular sites with lower carbon footprints, nor does it aim to set standards for carbon emissions. The recognition scheme uses carbon footprint information to provide information to the public about the potential general consequences of their travel, in an effort to increase consideration of climate effects.

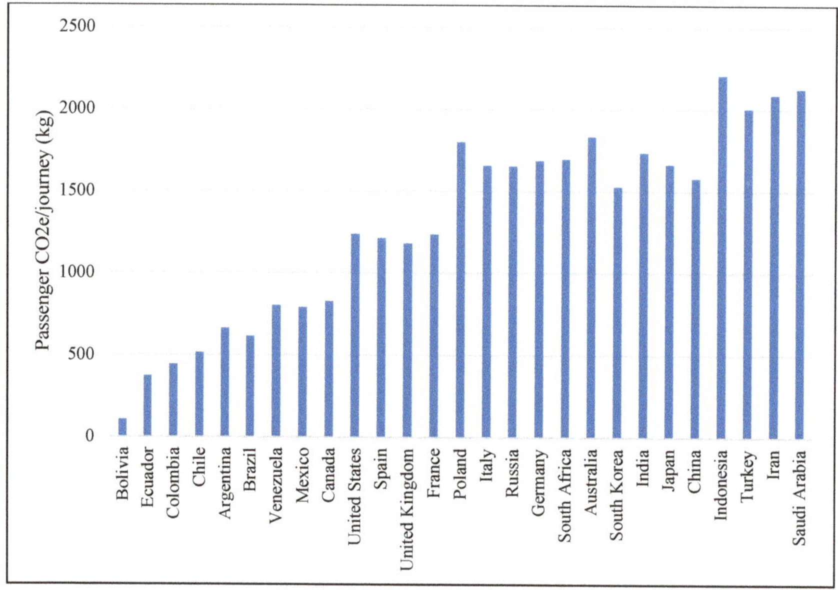

Figure 2. Carbon dioxide equivalent (CO2e) emissions generated in a round-trip flight from a representative airport city in each country to Cusco, Peru, for the purpose of visiting Machu Picchu.

2.3. Collecting Additional Climate Information

In order to address the narrative potential presented by sites in this subgroup for climate change communication, a list of a priori themes were developed that may be applied to coding site documentation. Documentation such as advisory body evaluations, media articles related to the site, site management plans, UNESCO mission reports, periodic reporting by the State Party, and other relevant reports were collected from the UNESCO website and other web sources. These were analyzed using NVivo 12 through a directed approach to content analysis [56]. As analysis proceeded, additional themes arose, and some themes were combined to clarify patterns. The final list of themes addressed topics such as conflict, education, environment, climate change impacts, biodiversity, industrial heritage, public engagement, sustainable development, and transnationalism. Patterns amongst the data were then drawn out and used as a framework for a brief narrative about how best to draw a visitor into a story about climate change at the site.

Identification of sustainability practices took place through reviews of site documentation, work conducted by UNESCO and other researchers, and news media. Sustainability measures considered include, for example, aggressive action towards increased use of renewable energy, acknowledgment of the importance of energy audits and reducing energy use, and embracing renewable energy at the beginning of electrification projects.

Climate change impacts were identified through reviews of site documentation, UNESCO reports, news media, and academic work. Vulnerabilities to climate change and previously-recorded impacts of climate change on sites were documented. These focused on the physical, direct impacts of climate change on the site. Damage caused by conflict related to climate change was not included in this section (but was considered for the narrative potential of sites).

3. Results

The data on carbon footprints, climate narratives, sustainability practices, and climate change impacts were imported into ArcGIS Online, a free, open-source, cloud-based mapping tool. The ArcGIS StoryMaps platform was then used to combine the text, images of sites, and interactive maps to create a publicly-accessible platform to present the results of the analysis, developed as a Climate Communication Recognition Scheme titled *Climate Footprints of Heritage Tourism* [41]. The StoryMap interface allows for the creation of interactive web-based platforms built around the presentation of spatial information through maps. By linking narrative text with photographs and maps of spatial relationships, StoryMaps are a rich, visually compelling way to present information that engages the reader and facilitates dissemination [57,58]. ArcGIS StoryMaps are increasingly being used across fields to communicate information, including in education, public health, museum studies, and geology [59–62].

The StoryMap developed in this project presents the data collected for each of the subgroups in two overarching themes: Understanding Your Impact and Telling the Story:

- Understanding Your Impact provides an opportunity to learn about the effects of travel to WHS through the carbon footprint analysis. This section includes a description of what carbon footprint analysis is and the methodology for this part of the project. It then provides the results of the carbon footprint analysis in narrative form for each of ten WHS (Figure 3). Each site description concludes with an interactive map that shows the carbon cost of international flights to the WHS from 1 of at least 20 countries, which represent the 20 countries with the greatest carbon emissions in 2015 [63] and the countries which send the most tourists to the WHS country.
- Telling the Story provides a hub of information about the ways climate change and World Heritage are related, through climate narratives, sustainability actions, and climate change impacts (Figure 4). This section of the platform highlights the WHS for each subgroup on a world map, including the site's spatial location, photographs, narrative text, and links to more information.

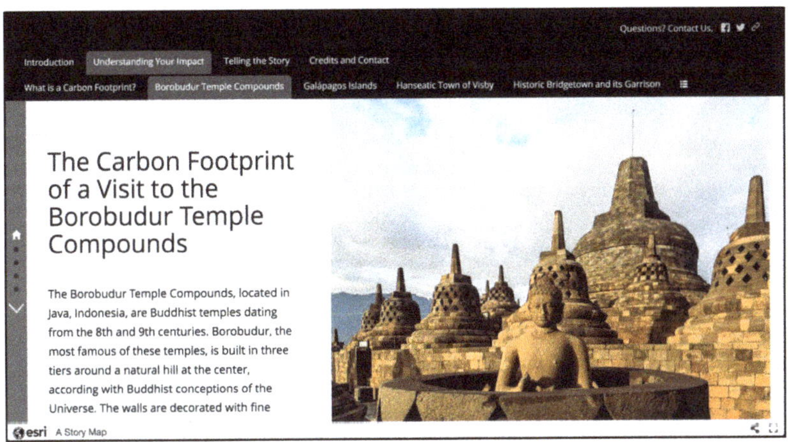

Figure 3. Webpage screenshot of Understanding Your Impact theme, with example of the site of the Borobudur Temple Compounds, in the StoryMap *Climate Footprints of Heritage Tourism* [41].

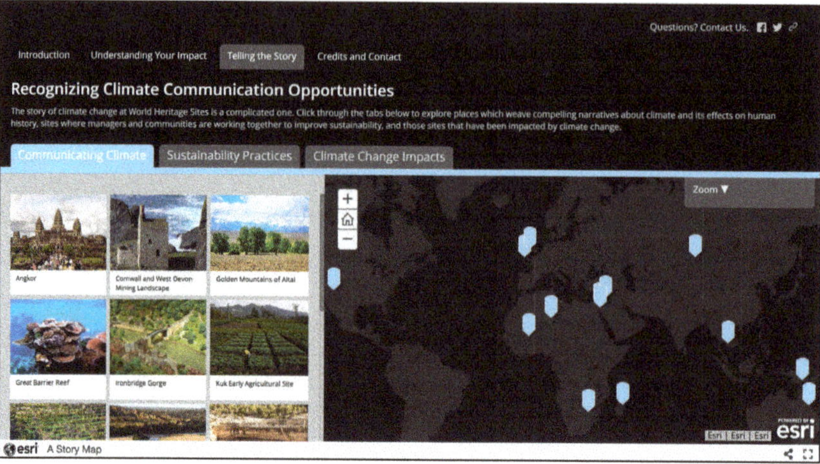

Figure 4. Webpage screenshot of Telling the Story theme in the StoryMap *Climate Footprints of Heritage Tourism* [41].

4. Discussion: Climate Communication Recognition Schemes as a Transnational Governance Tool

This study represents a pilot approach for developing a public-facing instrument that uses heritage sites to communicate various facets of climate change, including information relevant for mitigation and adaptation responses. Specifically, we developed the instrument as a CCRS to provide a tool for transnational governance. CCRS is an elaboration on the concept of ecolabels, but with two primary distinctions: a CCRS is specifically oriented towards climate concerns and captures a more holistic portrait, for example of the entanglements between heritage resources and climate change, than the strictly quantitative measures typically employed and available in ecolabels. We consider that the development of ecolabels and CCRS will strengthen the knowledge base for sustainable site management and consumer decision-making in the face of global climate change; most importantly, such instruments represent mechanisms for transnational governance of climate issues, in the context of the deficits and dysfunction in the international governance of climate change responses.

Increasingly issues or policy problems—like global climate change—that impact heritage resources and their management extend beyond state jurisdictions. Furthermore, they are poorly addressed or captured by the traditional international system of state-to-state treaties and treaty-based intergovernmental organizations, for example by UNESCO World Heritage and its system of conventions, recommendations, and associated advising bodies. International organizations like UNESCO are subject to the grindingly slow machinery of national and international governance, which have proven inadequate to address the rising environmental and social challenges of climate change. Additionally, some of the deficits in the ability of UNESCO heritage frameworks to respond in a meaningful way to climate change are baked into UNESCO's conventions. International agreements like the UNESCO conventions related to heritage (e.g., the 1972 Convention Concerning the Protection of the World Cultural and Natural Heritage and the 2003 Convention for the Safeguarding of the Intangible Cultural Heritage) make no explicit reference to climate change, though the World Heritage Convention does address threats to heritage sites presented by extreme weather [64]. Indeed, though scholars and climate activists have explored the extent to which the World Heritage Convention might be used to compel States Parties to take action on climate change [65,66], others reiterate that these conventions assert the rights of states to protect their own heritage sites, and do not enforce obligations to mitigate climate change [67]. Despite the shortcomings of UNESCO World Heritage as an international governmental body, we do find value in the branding function—as a brand signifying global conservation—that UNESCO World Heritage has come to take on, especially in the shift towards development models of heritage management. This branding function may be usefully developed as a transnational governance tool: as an ecolabel or, in this case, the development and refinement of an ecolabel as a CCRS.

Scholars and actors in climate governance are turning to transnational governance as one of the most promising arenas for climate action [68–70]. Given the limitations of the international system for cross-border governance and interstate cooperation, developing transnational governance tools (like ecolabels) offers the opportunity for institutional innovations that better fit heritage management practices to the challenges of global climate change. The search for institutional innovations in climate governance is the rationale for this study to develop a CCRS for WHS. More broadly, heritage resources and heritage practices increasingly cross national borders, underscoring the need for the further development of heritage management paradigms within transnational frameworks [39,71].

It is important to understand that, across various sectors and contexts, transnational governance has assumed similar functions as the state and the international system, such as rule-making, implementation, monitoring, enforcement, and service provision. However, transnational governance is different from state governments and international system on account of having shed the administrative and institutional trappings of these governmental bodies. This move from the state/international to transnational is perhaps best understood as a shift from "governments" to "governance." This shift has been ongoing for some time now, as structural adjustments to government under neoliberal reforms have encouraged the decentralization of state bureaucracies and withdrawal of the state from social provisioning.

In general, transnational governance is characterized by several key trends: (1) merging of domestic and international politics; (2) increasing involvement of non-state actors (e.g., individuals, firms, corporations, and NGOs) in global politics, in providing technical expertise, monitoring, lobbying, or enforcement; (3) private governance, e.g., the involvement of private foundations in global standards, such as the Gates Foundation in global health; and (4) new methods for eliciting compliance with transborder standards, such as 'naming and shaming,' capacity building, persuasion, transparency, and dissemination of best practices [72] (pp. 6–12). Overall, governance mechanisms for organizing action and eliciting compliance—e.g., through the private sphere, non-governmental organizations (NGOs), and corporate social responsibility (CSR)—are being sought beyond the international system.

Ecolabels fit squarely within the above trends, as they are typically developed by non-state actors, represent a form of private governance, and provide a tool for eliciting compliance. Ecolabeling uses a certification process to assign labels to consumer goods, e.g., organic and fair trade certification of food, for communicating the sustainability of a product [73–75]. Ecolabels extend to tourism; for example 30% (128 of 430) of the ecolabels listed in the 2014 Ecolabel Index apply to the tourism industry [74]. In the context of the tourism industry, these schemes often carry a strong quantifiable and ecological focus. However, efforts to incorporate cultural, social, and economic aspects of a tourist destination into these schemes are underway and are often reflected in a more holistic approach to certification processes known as an "environmental product information scheme" (EPIS) that aim to provide more holistic ecological information on products and services [75–78]. The emergence of voluntary transnational governance tools for encouraging and recognizing environmental sustainability has resulted in a variety of instruments that can collectively be called EPIS. These schemes include mechanisms such as ecolabels, stewardship certificates, green trademarks, rankings and ratings, codes of conduct, and standards for reports and declarations, among others [79,80]. Ecolabels and EPIS are employed to communicate information to consumers, such as quality and sustainability, and reward and promote goods and services that are environmentally superior in some respect [75]. They provide consumers with the opportunity to "vote with their wallet" in a manner of environmental policymaking that extends beyond national governments and borders.

The current study is situated within this literature and, as described above, aims to develop an environmental product information scheme for communicating climate change at WHS, which we have called a "climate communication recognition scheme" (CCRS). This elaboration from an ecolabel to a more holistic capturing of the various facets of climate change knowledge and responses immanent in heritage sites is reflected in our choices in designing the CCRS. In addition to the quantitative measures that typically characterize ecolabels, given in this case by carbon footprint analysis, we have also chosen to portray other facets of climate change at heritage sites, such as their narrative potential for communicating the histories of anthropogenic climate change, the sustainability practices being undertaken at heritage sites, and the impacts of climate change to heritage resources. We consider this approach a more accurate capturing of the climate communication potential of heritage sites, not only pursuing carbon management tools like carbon footprint analysis, but also triangulated with social justice and sustainability issues attendant with heritage sites and global climate change (for example, exacerbating existing social and economic inequalities, the fundamental and paradoxical inequalities of climate change between who contributes and who suffers, and the impacts of climate change for increasing migration, displacement, conflict, and food insecurity). WHS are iconic institutions representing the cultural riches and creative solutions that societies around the world have produced in relationship to their environment, so it is important that an ecolabel for WHS is situated within a holistic context of social, economic, and environmental sustainability and attentive to the social justice dimensions of global climate change.

One of the limitations of this study is the time-intensive nature of collecting information for each of the WHS included in the StoryMap. Given the sheer number of sites on the World Heritage List, expanding the approach taken here to the entire list would require significant time investment and is probably unrealistic. Plus, taking into account that new sites are inscribed to the World Heritage List each year, we suggest that any CCRS for WHS should be considered a "living" and provisional labeling tool rather than an exhaustive one. Therefore, future development of the CCRS beyond our pilot-study will need to set evaluative rubrics for which WHS should be prioritized over others for inclusion in the CCRS. The calculation of carbon footprints of travel to the sites is also particularly time-intensive, as it requires collecting considerable amounts of data from various online sources. A further limitation of the carbon footprint calculations, in addition to those discussed in the methods section above, is that these footprints were calculated under the assumption that the tourist was traveling only to the site in question. Generally, heritage tourism is part of a tourism circuit, in which visitors to a country travel to several locations on one trip [81].

Following this pilot-study, we plan to continue developing and refining the CCRS for eventual dissemination amongst heritage managers and specialists as well as the general public. In addition to increasing the number of sites represented in the CCRS, we also intend to continue building out the narrative descriptions for each site, especially to underscore the holistic nexus of climate impacts, adaptive capacity, carbon management and other mitigatory actions, and climate communication opportunities that exist, or have the potential to exist, at every site. A holistic approach to climate change in site management, heritage management, and public engagement with heritage sites will offer better opportunities and strengthen resilience for responding to the challenges of climate change. Furthermore, our work with the CCRS mirrors broader developments in formalizing the inclusion of climate considerations in heritage management, and though these trends are still primarily focused on impacts and adaptation, such analyses and instruments as provided here may be of interest in strengthening and institutionalizing climate change responses within heritage site management and heritage management more broadly.

Overall, we offer this study as an example of the development of transnational governance tools in heritage management for addressing climate change, and we recommend further research and initiatives in this direction. As we develop in this study, an ecolabeling scheme for WHS that communicates the holistic facets of climate change could influence consumer/tourist choices and site management for a well-known "brand" of tourism sites that trades on global awareness and global conservation. We urge further development of transnational tools for heritage management in order to meet the climate crisis. The ineffectiveness of climate action within the international system behooves stakeholders of heritage resources to pursue sustainable development and sustainable tourism via low-carbon economic growth and to innovate and incorporate transnational tools for responding to climate change, which will serve to "fill in the gaps" or provide supportive scaffolding to international and national heritage policies and practices.

Author Contributions: All authors helped in methodology, formal analysis, investigation, data curation, and developing the manuscript. Conceptualization, funding acquisition, and supervision, K.L.S. All authors have read and agreed to the published version of the manuscript.

Funding: This research was funded by the University of Maryland ADVANCE Program for Inclusive Excellence and the University of Maryland Department of Anthropology.

Acknowledgments: We wish to thank Klaus Hubacek for his advice during the conceptualization of this project, and Paul Shackel and Barnet Pavão-Zuckerman for their feedback on the development of the StoryMap. This paper benefitted from the insightful recommendations of five anonymous reviewers, and we thank them for their time and attention. We are also grateful for website support from the University of Maryland Department of Anthropology for the website http://www.heritageofclimate.com.

Conflicts of Interest: The authors declare no conflict of interest.

Appendix A

Subgroups of WHS chosen for climate communication recognition scheme *Climate Footprints of Heritage Tourism* (some sites are included in more than one subgroup):

(a) Carbon Footprint Analysis

1. Borobudur Temple Compounds
2. Galapágos Islands
3. Hanseatic Town of Visby
4. Historic Bridgetown and Garrison
5. Historic Sanctuary of Machu Picchu
6. Kakadu National Park
7. Masada
8. Nubian Monuments from Abu Simbel to Philae
9. Robben Island
10. Statue of Liberty

(b) Sites with Narrative Potential

11. Angkor
12. Cornwall and West Devon Mining Landscape
13. Golden Mountains of Altai
14. Great Barrier Reef
15. Ironbridge Gorge
16. Kuk Early Agricultural Site
17. Land of Olives and Vines
18. Mapungubwe Cultural Landscape
19. Mesa Verde National Park
20. Palmyra
21. Petra
22. Royal Hill of Ambohimanga
23. Tassili n'Ajjer
24. Timbuktu

(c) **Sites Pursuing Sustainability Practices**

25. Costiera Amalfitana
26. Ferrara, City of the Renaissance, and its Po Delta
27. Galapágos Islands
28. Golden Mountains of Altai
29. Great Barrier Reef
30. Hanseatic Town of Visby
31. Historic Centre of Oaxaca and Archaeological Site of Monte Albán
32. Loire Valley between Sully-sur-Loire and Chalonnes
33. Old and New Towns of Edinburgh
34. Petra
35. Rice Terraces of the Philippine Cordilleras
36. Robben Island
37. Routes of Santiago de Compostela (Camino de Santiago)
38. Taj Mahal

(d) **Sites Most Impacted by Climate Change (Current and Projected)**

39. Agave Landscape and Ancient Industrial Facilities of Tequila
40. Ancient Ksour of Ouadane, Chinguetti, Tichitt and Oualata
41. Borobudur Temple Compounds
42. Cilento and Vallo di Diano National Park with the Archaeological Sites of Paestum and Velia, and the Certosa di Padula
43. Coffee Cultural Landscape
44. Cornwall and West Devon Mining Landscape
45. Costiera Amalfitana
46. Ferrara, City of the Renaissance, and its Po Delta
47. Galapágos Islands
48. Golden Mountains of Altai
49. Great Barrier Reef
50. Historic Bridgetown and Garrison
51. Historic Sanctuary of Machu Picchu
52. Historical Complex of Split with the Palace of Diocletian
53. Hoi, An Ancient Town
54. Ibiza, Biodiversity and Culture
55. Kakadu National Park
56. Landscape of the Pico Island Vineyard Culture
57. Lavaux, Vineyard Terraces
58. Medina of Essaouria (Mogador)
59. Mesa Verde National Park
60. Ouadi Qadisha (the Holy Valley) and the Forest of the Cedars of God
61. Port, Fortresses and Group of Monuments, Cartagena
62. Rice Terraces of the Philippine Cordilleras
63. Robben Island
64. Ruins of Kilwa Kisiwani and Ruins of Songo Mnara
65. Saloum Delta
66. SGang Gwaay
67. Statue of Liberty

68. Timbuktu
69. Venice and its Lagoon
70. Viñales Valley

References

1. Gössling, S. *Carbon Management in Tourism: Mitigating the Impacts on Climate Change*; Routledge: New York, NY, USA, 2010.
2. Hall, C.M.; Scott, D.; Gössling, S. The primacy of climate change for sustainable international tourism. *Sustain. Dev.* **2013**, *21*, 112–121. [CrossRef]
3. Scott, D.; Hall, C.M.; Gössling, S. *Tourism and Climate Change: Impacts, Adaptation and Mitigation*; Routledge: New York, NY, USA, 2012.
4. Scott, D.; Gössling, S.; Hall, C.M. International tourism and climate change. *Wiley Interdiscip. Rev. Clim. Chang.* **2012**, *3*, 213–232. [CrossRef]
5. Hall, C.M. Heritage, heritage tourism and climate change. *J. Herit. Tour.* **2016**, *11*, 1–9. [CrossRef]
6. UNWTO (United Nations World Tourism Organization), UNEP (United Nations Environment Programme), and WMO (World Meteorological Organization). *Climate Change and Tourism: Responding to Global Challenges*; UNWTO/UNEP/WMO: Madrid, Spain, 2008.
7. WEF (World Economic Forum). *Towards a Low Carbon Travel and Tourism Sector*; World Economic Forum: Davos, Switzerland, 2009.
8. Lenzen, M.; Sun, Y.-Y.; Faturay, F.; Ting, Y.-P.; Geschke, A.; Malik, A. The carbon footprint of global tourism. *Nat. Clim. Change* **2018**, *8*, 522–528. [CrossRef]
9. International Air Transport Association (IATA). IATA Forecast Predicts 8.2 Billion Air Travelers in 2037 [Summary of IATA 20-year Passenger Forecast]. 2018. Available online: http://www.iata.org/pressroom/pr/Pages/2018-10-24-02.aspx (accessed on 31 October 2019).
10. Gössling, S.; Hall, C.M. *Tourism and Global Ecological Change: Ecological, Social, Economic and Political Interrelationships*; Routledge: New York, NY, USA, 2006.
11. Becken, S.; Hay, J. *Climate Change and Tourism: From Policy to Practice*; Routledge: New York, NY, USA, 2012.
12. Kuo, N.-W.; Lin, C.-Y.; Chen, P.-H.; Chen, Y.-W. An inventory of the energy use and carbon dioxide emissions from island tourism based on a life cycle assessment approach. *Environ. Prog. Sustain. Energy* **2012**, *31*, 3–465. [CrossRef]
13. Filimonau, V.; Dickinson, J.; Robb ins, D.; Reddy, M.V. The role of 'indirect' greenhouse gas emissions in tourism: Assessing the hidden carbon impacts from a holiday package tour. *Trans. Res. Part A Policy Pract.* **2013**, *54*, 78–91. [CrossRef]
14. Hall, C.M.; Baird, T.; James, M.; Ram, Y. Climate change and cultural heritage: Conservation and heritage tourism in the Anthropocene. *J. Herit. Tour.* **2016**, *11*, 1–24. [CrossRef]
15. Hansen, J. *Storms of My Grandchildren: The Truth About the Coming Climate Catastrophe and Our Last Chance to Save Humanity*; Bloomsbury: New York, NY, USA, 2009.
16. Beniston, M. Sustainability of the landscape of a UNESCO World Heritage Site in the Lake Geneva Region (Switzerland) in a greenhouse climate. *Int. J. Climatol.* **2007**, *28*, 1519–1524. [CrossRef]
17. Burns, W.C.G. Belt and suspenders? The World Heritage Convention's role in confronting climate change. *Rev. Eur. Commun. Int. Environ. Law* **2009**, *18*, 148–163. [CrossRef]
18. Burt, S.; Boom, K.; Lawrence, J.; Leonard, S.; Rosenthal, E.; Wagner, J.M. *The Role of Black Carbon in Endangering World Herit. Sites Threatened by Glacial Melt and Sea Level Rise*; Earthjustice and Australian Climate Justice Program: Oakland, CA, USA; Ultimo, Australia, 2009.
19. Dawson, J.; Johnston, M.J.; Stewart, E.J.; Lemieux, C.J.; Lemelin, R.H.; Maher, P.T.; Grimwood, B.S.R. Ethical Considerations of Last Chance Tourism. *J. Ecotour.* **2011**, *10*, 250–265. [CrossRef]
20. Frew, E.A. Advertising World Heritage Sites: Tour Operators and Last Chance Destinations. In *Last Chance Tourism: Adapting Tour. Opportunities in a Changing World*; Lemelin, R.H., Dawson, J., Stewart, E.J., Eds.; Routledge: New York, NY, USA, 2012; pp. 117–132.
21. Lemelin, H.; Dawson, J.; Stewart, E.J.; Maher, P.; Lueck, M. Last-chance tourism: The boom, doom, and gloom of visiting vanishing destinations. *Curr. Issues Tour.* **2010**, *13*, 477–493. [CrossRef]

22. Lemelin, H.R.; Dawson, J.; Stewart, E.J. *Last Chance Tour.: Adapting Tour. Opportunities in a Changing World*; Routledge: Abingdon, UK, 2012.

23. Eijgelaar, E.; Thaper, C.; Peeters, P. Antarctic cruise tourism: The paradoxes of ambassadorship, 'last chance tourism,' and greenhouse gas emissions. *J. Sustain. Tour.* **2010**, *18*, 337–354. [CrossRef]

24. Dawson, J.; Stewart, E.J.; Lemelin, H.; Scott, D. The carbon cost of polar bear viewing tourism in Churchill, Canada. *J. Sustain. Tour.* **2010**, *18*, 319–336. [CrossRef]

25. Buckley, R. The effects of World Heritage listing on tourism to Australian national parks. *J. Sustain. Tour.* **2004**, *12*, 70–84. [CrossRef]

26. PricewaterhouseCoopers. *The Costs and Benefits of UK World Heritage Site Status*; UK Department for Culture, Media and Sport: London, UK, 2007.

27. Rebanks Consulting, and Trends Business Research. World Heritage Status: Is There Opportunity for Economic Gain. In *Research and Analysis of the Socio-Economic Impact Potential of UNESCO World Heritage Site Status*; Lake District World Heritage Project: Kendal, UK, 2009.

28. Su, Y.-W.; Lin, H.-L. Analysis of international tourist arrivals worldwide: The role of World Heritage Sites. *Tour. Manag.* **2014**, *40*, 46–58. [CrossRef]

29. Yang, C.-H.; Lin, H.-L.; Han, C.-C. Analysis of international tourist arrivals in China: The role of World Heritage Sites. *Tour. Manag.* **2010**, *31*, 827–837. [CrossRef]

30. Adie, B.A. Franchising our heritage: The UNESCO World Heritage brand. *Tour. Manag. Perspect.* **2017**, *24*, 48–53. [CrossRef]

31. King, L.M.; Halpenny, E.A. Communicating the World Heritage brand: Visitor awareness of UNESCO's World Heritage symbol and the implications for sites, stakeholders and sustainable management. *J. Sustain. Tour.* **2014**, *22*, 768–786. [CrossRef]

32. Ryan, J.; Silvanto, S. A study of the key strategic drivers of the use of the World Heritage Site designation as a destination brand. *J. Travel Tour. Mark.* **2014**, *31*, 327–343. [CrossRef]

33. Wuepper, D.; Patry, M. The World Heritage List: Which sites promote the brand? A big data spatial econometrics approach. *J. Cult. Econ.* **2017**, *41*, 1–21. [CrossRef]

34. Drost, A. Developing sustainable tourism for World Heritage Sites. *Annal. Tour. Res.* **1996**, *23*, 479–484. [CrossRef]

35. Landorf, C. Managing for sustainable tourism: A review of six cultural World Heritage Sites. *J. Sustain. Tour.* **2009**, *17*, 53–70. [CrossRef]

36. Schmutz, V.; Elliott, M.A. Tourism and sustainability in the evaluation of World Heritage Sites, 1980–2010. *Sustainability* **2016**, *8*, 261–275. [CrossRef]

37. Nzama, A.T. The nexus between sustainable livelihoods and ecological management of the World Heritage Sites: Lessons from iSimangaliso World Heritage Park, South Africa. *Inkanyiso J. Hum. Soc. Sci.* **2009**, *1*, 34–42. [CrossRef]

38. Li, M.; Wu, B.; Cai, L. Tourism development of World Heritage Sites in China: A geographic perspective. *Tour. Manag.* **2008**, *29*, 308–319. [CrossRef]

39. Lafrenz Samuels, K. Transnational turns for archaeological heritage: From conservation to development, governments to governance. *J. Field Archaeol.* **2016**, *41*, 355–367. [CrossRef]

40. Lafrenz Samuels, K.; Lilley, I. Transnationalism and Heritage Development. In *Global Heritage: A Reader*; Meskell, L., Ed.; Blackwell: Malden, MA, USA, 2015; pp. 217–239.

41. Climate Footprints of Heritage Tourism. Available online: http://heritageofclimate.com (accessed on 4 March 2020).

42. Markham, A.; Osipova, E.; Lafrenz Samuels, K.; Caldas, A. *World Heritage and Tour. in a Changing Climate*; United Nations Environment Programme (UNEP) & United Nations Educational, Scientific and Cultural Organization (UNESCO): Nairobi, Kenya; Paris, France, 2016.

43. Marzeion, B.; Levermann, A. Loss of cultural World Heritage and currently inhabited places to sea-level rise. *Environ. Res. Lett.* **2014**, *9*, 3–7. [CrossRef]

44. United Nations Educational, Scientific and Cultural Organization (UNESCO). *Case Studies on Climate Change and World Heritage*; UNESCO: Paris, France, 2007.

45. United Nations Educational, Scientific and Cultural Organization (UNESCO). *Policy Document on the Impacts of Climate Change on World Heritage Properties*; UNESCO: Paris, France, 2008.

46. Fløttum, K.; Gjerstad, Ø. Narratives in climate change discourse. *Wiley Interdiscip. Rev. Clim. Change* **2017**, *8*, e429. [CrossRef]

47. Coulter, L.; Serrao-Neumann, S.; Coiacetto, E. Climate Change adaptation narratives: Linking climate knowledge and future thinking. *Futures* **2019**, *111*, 57–70. [CrossRef]

48. Saldaña, J. *The Coding Manual for Qualitative Researchers*; SAGE Publications: London, UK, 2009.

49. Available online: http://whc.unesco.org/en/list/ (accessed on 4 March 2020).

50. Benchikh, O.; Cipriano, M. *Good Practices: Success Stories on Sustainable and Renewable Energies in UNESCO Sites*; UNESCO: Paris, France, 2013.

51. Booysen, J. Robben Island Is Going Solar. IOL Business Report. 18 May 2018. Available online: http://www.iol.co.za/business-report/energy/robben-island-is-going-solar-15029020 (accessed on 4 November 2019).

52. El Hanandeh, A. Quantifying the carbon footprint of religious tourism: The case of Hajj. *J. Clean. Prod.* **2013**, *52*, 53–60. [CrossRef]

53. Sharp, H.; Grundius, J.; Heinonen, J. Carbon footprint of inbound tourism to Iceland: A consumption-based life-cycle assessment including direct and indirect emissions. *Sustainability* **2016**, *8*, 11. [CrossRef]

54. United States Environmental Protection Agency Greenhouse Gas Equivalencies Calculator. Available online: http://www.epa.gov/energy/greenhouse-gas-equivalencies-calculator (accessed on 4 March 2020).

55. International Civil Aviation Organization (ICAO) Carbon Emissions Calculator. Available online: http://www.icao.int/environmental-protection/Carbonoffset/Pages/default.aspx (accessed on 4 March 2020).

56. Hsieh, H.-F.; Shannon, S.E. Three approaches to qualitative content analysis. *Qual. Health Res.* **2005**, *15*, 1277–1288. [CrossRef] [PubMed]

57. Molden, O.C. Short take: Story-mapping experiences. *Field Methods* **2019**, 1–9. [CrossRef]

58. West, H.; Horswell, M. GIS has changed! Exploring the potential of ArcGIS online. *Teach. Geogr.* **2018**, *43*, 22–24.

59. Walshe, N. Using ArcGIS online story maps. *Teach. Geogr.* **2016**, *41*, 115–117.

60. Rauch, S. Mapping PPS: A case study of story map journals for interactive health reporting. *Online J. Public Health Inf.* **2019**, *11*, e313. [CrossRef]

61. Hunt, R.A. Accessibility of Museum Collections through ArcGIS Story Maps at Lake Champlain Maritime Museum. Master's Thesis, State University of New York at Binghamton, New York, NY, USA, 2019.

62. Antoniou, V.; Ragia, L.; Nomikou, P.; Bardouli, P.; Lampridou, D.; Ioannou, T.; Kalisperakis, I.; Stentoumis, C. Creating a story map Using Geographic Information Systems to explore geomorphology and history of Methana Peninsula. *ISPRS Int. J. Geo-Inf.* **2018**, *7*, 12. [CrossRef]

63. Union of Concerned Scientists. Each Country's Share of CO2 Emissions (10 October 2018). Available online: http://www.ucsusa.org/global-warming/science-and-impacts/science/each-countrys-share-of-co2.html (accessed on 4 November 2019).

64. Carducci, G. What Consideration Is Given to Climate and to Climate Change in the UNESCO Cultural Heritage and Property Conventions? In *Climate Change as a Threat to Peace: Impacts on Cultural Heritage and Cultural Diversity*; von Schorlemer, S., Maus, S., Eds.; PL Academic Research: Frankfurt am Main, Germany, 2014; pp. 129–139. Available online: http://www.jstor.org/stable/j.ctv2t4cvp (accessed on 4 March 2020).

65. Chechi, A. The Cultural Dimension of Climate Change: Some Remarks on the Interface Between Cultural Heritage and Climate Change Law. In *Climate Change as a Threat to Peace: Impacts on Cultural Heritage and Cultural Diversity*; von Schorlemer, S., Maus, S., Eds.; PL Academic Research: Frankfurt am Main, Germany, 2014; pp. 161–197. Available online: http://www.jstor.org/stable/j.ctv2t4cvp (accessed on 4 March 2020).

66. Quirico, O. Key Issues in the Relationship Between the World Heritage Convention and Climate Change Regulation. In *Cultural Heritage, Cultural Rights, Cultural Diversity: New Developments in International Law*; Borelli, S., Lenzerini, F., Eds.; Studies in Intercultural Human Rights 4. Martinus Nijhoff Publishers: Leiden, The Netherlands, 2012; pp. 291–412.

67. Lenzerini, F. Protecting the Tangible, Safeguarding the Intangible: A Same Conventional Model for Different Needs. In *Climate Change as a Threat to Peace: Impacts on Cultural Heritage and Cultural Diversity*; von Schorlemer, S., Maus, S., Eds.; PL Academic Research: Frankfurt am Main, Germany, 2014; pp. 141–160. Available online: https://www.jstor.org/stable/j.ctv2t4cvp (accessed on 4 March 2020).

68. Bulkeley, H.; Andonova, L.B.; Betsill, M.M.; Compagnon, D.; Hale, T.; Hoffmann, M.J.; Newell, P.; Paterson, M.; VanDeveer, S.D.; Roger, C. *Transnational Climate Change Governance*; Cambridge University Press: Cambridge, UK, 2014.

69. Andonova, L.B.; Hale, T.N.; Roger, C.B. *The Comparative Politics of Transnational Climate Governance*; Routledge: New York, NY, USA, 2019.

70. Hickmann, T. *Rethinking Authority in Global Climate Governance: How Transnational Climate Initiatives Relate to the International Climate Regime*; Routledge: New York, NY, USA, 2015.

71. Lafrenz Samuels, K. *Mobilizing Heritage: Anthropological Practice and Transnational Prospects*; University Press of Florida: Gainesville, FL, USA, 2018.

72. Hale, T.; Held, D. *Handbook of Transnational Governance: Institutions and Innovations*; Polity Press: Cambridge, UK, 2011.

73. Bolwig, S.; Gibbon, P. *Emerging Product Carbon Footprint Standards and Schemes and their Possible Trade Impacts*; Risø National Laboratory for Sustainable Energy, Technical University of Denmark: Roskilde, Denmark, 2009.

74. Gössling, S.; Buckley, R. Carbon labels in tourism: Persuasive communication? *J. Clean. Prod.* **2016**, *111*, 358–369. [CrossRef]

75. Rubik, F.; Frankl, P. *The Future of Eco-Labelling: Making Environmental Product Information Systems Effective*; Greenleaf Publishing: Sheffield, UK, 2005.

76. Das, M.; Chatterjee, B. Ecotourism: A panacea or a predicament? *Tour. Manag. Perspect.* **2015**, *14*, 3–16. [CrossRef]

77. Melissen, F.; Koens, K.; Brinkman, M.; Smit, B. Sustainable development in the accommodation sector: A social dilemma perspective. *Tour. Manag. Perspect.* **2016**, *20*, 141–150. [CrossRef]

78. Tepelus, C.M.; Córdoba, R.C. Recognition schemes in tourism—From 'eco' to 'sustainability'? *J. Clean. Prod.* **2005**, *13*, 135–140. [CrossRef]

79. Scheer, D.; Rubik, F. Environmental Product Information Schemes: An overview. In *The Future of Eco-Labelling: Making Environmental Product Information Systems Effective*; Rubik, F., Frankl, P., Eds.; Greenleaf Publishing: Sheffield, UK, 2005; pp. 46–88.

80. Boström, M.; Klintman, M. *Eco-Standards, Product Labelling and Green Consumerism*; Palgrave Macmillan: London, UK, 2008.

81. Alberti, F.G.; Giusti, J.D. Cultural heritage, tourism and regional competitiveness: The Motor Valley Cluster. *City Cult. Soc.* **2012**, *3*, 261–273. [CrossRef]

Article

Preservation of Distemper Painting: Indoor Monitoring Tools for Risk Assessment and Decision Making in Kvernes Stave Church

Tone Marie Olstad [1], Anne Apalnes Ørnhøi [1], Nina Kjølsen Jernæs [1], Lavinia de Ferri [2], Ashley Freeman [2] and Chiara Bertolin [2,*]

1   Norwegian Institute for Cultural Heritage Research (NIKU), 0155 Oslo, Norway; tone.olstad@niku.no (T.M.O.); anne.ornhoi@niku.no (A.A.Ø.); nina.k.jernaes@niku.no (N.K.J.)
2   Department of Mechanical and Industrial Engineering, Norwegian University of Science and Technology, 7491 Trondheim, Norway; lavinia.de.ferri@ntnu.no (L.d.F.); ashley.a.freeman@ntnu.no (A.F.)
*   Correspondence: Chiara.bertolin@ntnu.no

Received: 15 January 2020; Accepted: 12 February 2020; Published: 14 February 2020

Abstract: During the Medieval period, over 1000 stave churches were thought to have been constructed in Norway. However, currently, only 28 of these churches remain and only 19 still have distemper wall paintings. The cultural significance of these structures, and more specifically their elaborate distemper wall paintings, has changed over time, as have the means and methods for preserving these monuments. Deeper knowledge of the current state of these structures, along with environmental monitoring and modeling will open the way to a better understanding of preservation. This paper presents a case study for unheated Norwegian wooden churches based on data collected from Kvernes stave church. There are three aims for this paper: (i) to describe the typical indoor conditions similar to the historic climate of stave churches; (ii) determine the common characteristics of distemper paint found within stave churches; (iii) and develop a risk assessment tool to evaluate the climate-induced risk factors in stave churches. The outcome of this work will contribute to research performed within the Sustainable Management of Heritage Buildings in a Long-term Perspective (SyMBoL) project which aims to develop a better understanding of climate induced risks for stave churches, and ultimately to better manage environmental risk.

Keywords: indoor climate; climate-induced risk; distempered paint; decision making; consolidation; monitoring; stave church

---

1. Introduction

The indoor microclimate of cultural heritage sites, especially churches which have forgone any type of comfort heating system, are directly governed by the external climate. The climate affects the structure based on the structure's dimension, internal partition, and building materials [1]. In historic churches, high relative humidity (RH) and low temperature (T) values are often found [2,3], making the indoor environment uncomfortable for visitors and patrons. To improve comfort for users, many sites installed heating systems in the 20th century [4]. However, this was not the case for Norwegian churches which started to install heating systems in the 19th century; some even well before 1897 when a law imposed the mandatory heating of churches. Newer heating systems were installed in most churches during the 20th century [5].

Many Norwegian churches are currently heated using two different practices: the churches are either heated only during the winter months, and left unheated during the summer, or they are only heated during services. The main consequence of these two practices is induced variations in the church's indoor climate (i.e., relative humidity as well as temperature) [4]. However, this is not the case for Kvernes stave church which has been maintained in its original condition since its construction and is still without a heating system. A possible reason why a heating system was never installed in Kvernes is because a new church was erected in 1893, replacing the original stave church which was no longer in ordinary use.

In the last decade or so, the effects of climate-induced change have become more and more evident in many scientific fields, including the cultural heritage field. Quick and erratic climatic variations are creating new conservation challenges. Challenges related to effective risk assessment of sites may include long-term monitoring plans coupled with analytical campaigns that are able to concurrently detect the conservation state of all the materials existing in a historic building. These risk assessments, recently in association with tools allowing the simulation of current or future scenarios, permit the creation of the most appropriate strategies for the management of monuments [6]. Conventional museums and archives' environmental set points are hard to follow when approaching the management of temperature and relative humidity conditions of historic churches [4], including unique wooden structures such as stave churches. There are key parameters which need to be taken into consideration when proposing a strategy to control indoor microclimate and its risk assessment. For stave churches these parameters include the construction methods and material, the decorative parts of the interior, the protocol for use of a heating system in churches as well possible conflicts when installing heating, ventilation and conditioning (HVAC) systems to control the indoor climate. These parameters are the foundation for conservation and adaptation strategies, with regards to predicting the effects of indoor climate change.

The primary aim of this paper is to present the typical indoor conditions resembling historic climate information for stave churches. In particular, this paper will use Kvernes stave church as a case study for unheated Norwegian wooden churches, since it is preserved in its natural microclimate and not perturbed by a heating system. Secondly, this paper intends to present the characteristics of one of the most climate-sensitive materials preserved within a stave church—distemper paint—and more precisely distemper paint in Kvernes stave church which is not changed or "distorted" by overpainting, removal of overpaint or unsuitable consolidation treatments. In general, it is assessed to have been left untouched until certain painted areas on the walls were consolidated with sturgeon glue in 2013. Finally, this paper aims to introduce an effective standardized risk assessment tool [7] to the cultural heritage field to better manage the climate-induced decay of this sensitive object.

The research presented in this paper is a preliminary outcome of the International Research Project SyMBoL or Sustainable Management of Heritage Building in a Long-term Perspective (2018–2021) coordinated by the Norwegian University of Science and Technology (NTNU) and funded by the Research Council of Norway. The SyMBoL project aims to contribute to the debate on appropriate environmental conditions for preserving the 28 remaining Norwegian stave churches and their distemper decorative paint in a time of climate change and mass tourism.

1.1. Stave Churches and Distemper Decorative Paint

Distemper paintings are found in 19 of the 28 preserved stave churches in Norway. Most of the paintings are from the 17th and 18th centuries and are typically dominated by tendrils and vines, which often cover the entirety of the churches' interior. Seven of these 19 churches have distemper paintings, or fragments of paintings, from the 1200s (Torpo, Rollag, Nore, Hopperstad, Hedalen, Heddal, and the Høyjord; only the paintings in the Torpo stave churches are visible to the visitors) [8] (p. 69). Over the centuries, the artistic and cultural importance of these paintings have varied. Specifically, at the end of the 19th century, some of these decorative paint layers were deliberately washed off the church walls [8] (p. 71). However today these paintings are given high cultural and historic value, and the conservation of these paintings are now an integral part of conserving stave churches.

1.2. Kvernes Stave Church

Kvernes stave church is located on the west coast of Norway, in a coastal climate characterized by high relative humidity throughout the year and mild winters. The church is in Møre and Romsdal county, which happens to be one of Norway's top two counties most often impacted by extreme stormy weather [9]. The church is situated in a clearing, about 200 m from the fjords, completely exposed to the elements (Figure 1). After a hurricane in 1992, structural work was needed to prevent horizontal movement. In addition, the church has had continuous problems with water and moisture, both due to a leaky roof and moisture in the ground. As a result of these moisture issues, wood rot is a concern. In the hopes mitigating wood rot, the terrain around the church was lowered in the 1930s, reducing the amount of groundwater which could damage the wooden structure [10]. Even with these preventative measures, when examining the predicted climate towards the year 2100, there is an expected increase in average temperature of 2.0–2.5 °C, and the amount of rain is expected to increase by 25%–30%. In addition, more wind is also anticipated [10] (p. 13).

Figure 1. The exterior of Kvernes stave church during one of the rare days with snow (date: 3 March 2011). Photo © Norwegian Institute for Cultural Heritage Research (NIKU) 2011.

Kvernes is a single-nave church owned by the Society for the Preservation of Ancient Norwegian Monuments. It is nominally 16 m long and 7.5 m wide and seats up to 200 people (Figure 2). New research, carried out in 2019, might even date the structure of the church to as early as the 1600s [11] (p. 22). The entire exterior, except for the eastern wall, was cladded most likely during the mid-17th century. The interior, to date, is dominated by decorative paintings from the 1630s (Figure 3). These decorative distemper paintings completely cover the walls and ceiling in both the chancel and nave (Figure 3b,c), as well as the baptistery (Figure 3a) in the western end of the nave. Between the nave and the chancel is a screen composed of a dado (i.e., chair rail) under a row of turned balusters carrying horizontal moldings in two layers. King Christian VI's carved and painted monogram is centrally placed over the opening of the chancel. On each side of King Christian VI's monogram, one will find carved and painted coats of arms and a polychrome crucifix (ca. 1630s) hanging above the monogram. Connected with the screen, on the south wall, there is a carved and painted pulpit (ca. 1630s). In the chancel, a large, carved and painted polychrome epitaph (1671) dominates the northern wall. Resting on top of the wooden painted altar, is an altarpiece with a late Medieval triptych built into a carved and painted 17th century altarpiece (1695). On the nave's southwest wall, there is a small decorated gallery for the priest's family.

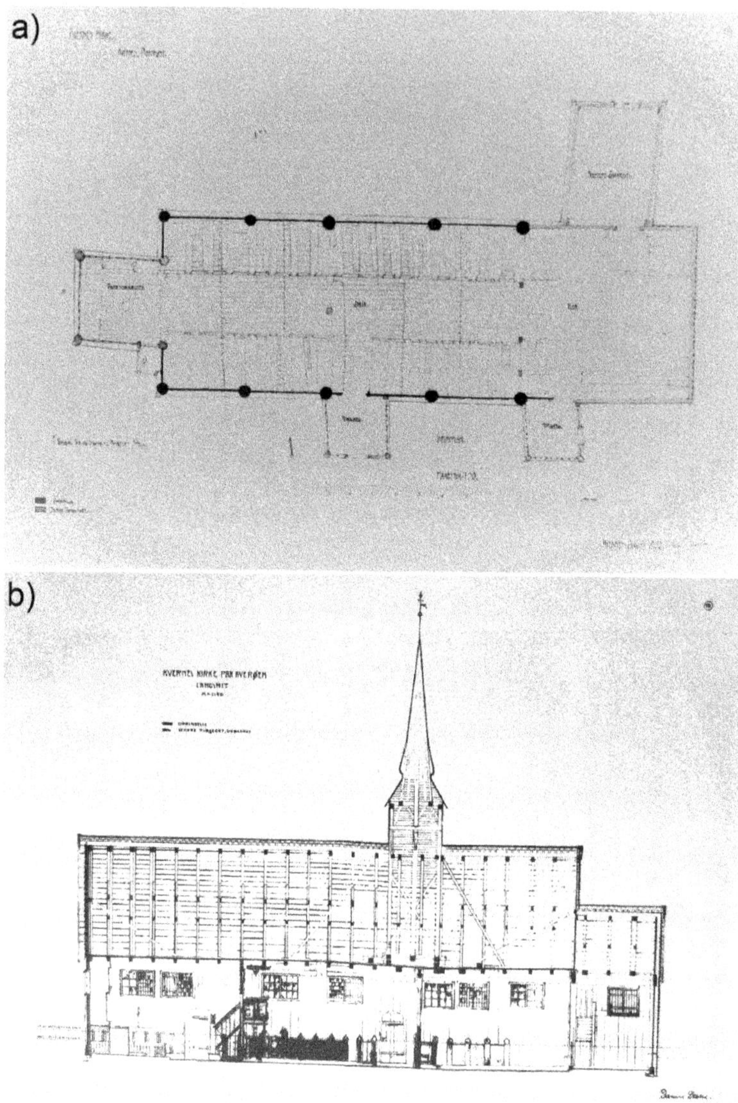

Figure 2. Kvernes stave church. (**a**) Planimetry; (**b**) vertical section. The south wall. Source: Illustrated guide to the stave church by the architect Heinrich Joachim Sebastian Karsten (1873–1947) in 1900.

Figure 3. Overview of Kvernes stave church. Photos © NIKU 2011. (**a**) Baptistery, towards west; (**b**) nave looking west; (**c**) nave looking east; (**d**) north wall of nave (western part); (**e**) south wall of nave (eastern part); (**f**) south wall of nave (central part). Date: 3 March 2011.

1.3. Distemper Paintings: Its Characteristics and Treatment

Distemper paint is a water-based material made from pigments mixed with animal glue as the main binder. However, analyses of 17th century paintings revealed that casein, egg, and oil are sometimes added to the glue binder [12]. Analyses carried out at the end of the 1990s identified casein or egg, however, this has yet to be confirmed by later studies. Furthermore, analyses performed as part of the Stave church preservation program, detected oil in some decorations dating back to the 1600s. It is uncertain whether this is due to a later treatment, or if oil was added to the glue as a binder. Analyzing glue as a protein-containing binder is complicated. The protein composition is specific for each type of glue and the proteins are defined by the combination of amino acids for each individual protein. In spite of several analysis methods used, in parallel, the chances of incorrect interpretation are high, and the amount of glue may be too small for the analysis instruments to detect.

The portions of binder to pigment in the distemper paint must be prepared in such a way that the paint neither runs down the wall whilst still wet, nor rubs off after drying. Distemper paintings are characterized by a porous, usually matte, surface.

Artists' exact process for painting decorative wall paintings in stave churches is unknown, however, through observation, analysis, and literature review, key concepts of artists' technique are elucidated. The painters did not put much effort into having the woodworkers prepare a smooth wooden surface to paint upon, as tool marks are often seen through the paint layers. However, the surface may have been primed with glue before the base paint was applied. A sketch on the base layer of paint, and the use of a compass and etched lines were observed in several distemper paintings from the 1600s. In post-Reformation distemper decorative paintings, the distinct local colors of the elements are painted on the surface of the dry background color, which was usually white; then any shadowing or colors that add shape, and finally the contours were painted. Color details could be added after the contouring. The paintings in the nave and baptistery in Kvernes stave church follow this general painting method (Figure 4).

Figure 4. (**a**) Close up of the distemper decorative paint in the nave in Kvernes stave church (date: 9 April 2013) Photo © NIKU 2013; (**b**) detail from the decorative pattern of the Urnes stave church (date: 30 April 2011) Photo © NIKU 2011; (**c**) details of the locations of Kvernes and Urnes stave churches with values of latitude and longitude.

Due to the matte and porous surface of distemper paint, very few methods and materials are suitable for conserving these paintings, if the main goal is to keep the appearance of the surface unchanged [13]. In 1984 the Directorate for Cultural Heritage (DCH) initiated a registration of distemper decorative paints in stave churches [14]. Subsequently, in 1989, the testing of different consolidation methods for distemper decorative paints was undertaken during the conservation treatment of the décor in Uvdal stave church (latitude: 60.26° N; longitude: 8.83° E) [15]. This test favored sturgeon glue, over many others, including, Klucel (hydroxypropylcellulose) and gelatin [16]. The conservators at the DCH, and later conservators at the Norwegian Institute for Cultural Heritage Research (NIKU), have since used sturgeon glue as the main consolidant for distemper paint [17]. The last consolidation treatment executed with sturgeon glue, in a Norwegian stave church, was in 2013 at Kvernes stave church. The preservation of distemper paint in churches is a constant area of interest for NIKU, and projects have concentrated on 16th and 18th century distemper decorative paintings [18]. Additional information was gained when the Stave Church Preservation Programme (2001–2015) provided an opportunity to work with several distemper decorative wall paintings that needed treatment: three Medieval distemper decorative paintings and 12 post-Reformation distemper decorative paintings in 14 stave churches were examined and treated [8].

An assessment of these treated paintings was made in 2013-2014. In several cases, the paint was found to lack proper adhesion to the underlying support; this was also found for those which were treated just a few years earlier. The examination was a follow-up of the assessment of consolidation methods for distemper decorative paints, as an answer to the constant need for greater knowledge on sturgeon glue as a consolidation material, and more specifically, for understanding which parameters are essential for successful results in consolidation treatments. In addition, the survey of conservative conditions of decorative distemper painted surfaces was also implemented in another recent project: "Environmental monitoring of the impact of climate change on protected buildings" started in 2017 and coordinated by NIKU [6]. An example of the mapping of conservation treatments obtained by the survey inside Kvernes stave church is depicted in Figure 5.

Figure 5. Example of mapping of conservation treatment in Kvernes stave church. The image shows the eastern part of the north wall in the nave of Kvernes stave church. Areas marked with green were consolidated with sturgeon glue (date: 5 April 2013). Photo © NIKU 2013.

2. Materials and Methods

2.1. Conservation Investigation of Distemper Paintings

In Section 2, the preservation state of the distemper paintings in the nave and baptistery of Kvernes is related to research and practical experience from the conservation of distemper paintings in a number of other stave churches. However, because the paintings in the chancel of Kvernes are so heavily altered by overpainting and conservation treatments, these paintings are not discussed in this paper. As for the nave, there are no indications of earlier treatment carried out on the distemper paintings. In general, signs of damage are related to wear and tear and long-term water leakages. The paintings are worn and unbound, and flaking paint was registered in 2012; however, the condition of the walls varies from area to area. This paper will only focus on distemper paint which was not heavily altered by previous conservation treatments.

In the case of Kvernes today, there are between 12,400 and 15,300 visitors a year, with an increasing number in the last four years. Usually two tourist buses arrive daily in the period April–September, in addition to tourists who travel by cruise in the period June–August. During these mass visits, the church administration only allows a maximum group of 90 visitors inside the church at a time (amount of visitor in the years 2016–2019, internal communication with Sørvik, A.K. Head of management Kvernes stave church. Email correspondence, 2 December 2019).

When assessing the current condition of the decorative paintings in the nave, during the last survey, the north wall was identified as being in the least ideal condition, with the paintings displaying different conservation conditions in different areas of the wall surface. At bench height, in the area were a staircase once existed and under the priest family's chair in the south-west corner, the paint has worn off with little paint left. In general, these areas reflect the degradation of distemper paint resulting from people rubbing the painted surface with their hands and coats [19] (pp. 42, 48). Further, the black color in the distemper paintings on the north wall was identified as having moisture and water damage [19] (p. 45). In the nave's northern part of the western wall, it can rain or snow through the wall if precipitation comes together with wind. In general, the building is not tight, so birds can sometimes find their way inside [19] (p. 48). In the ceiling of the nave, there are secondary beams, probably due to long-term water leakages [19] (p. 44). On the southern wall, most of the distemper decorative paint is damaged by water leakage, both from the ceiling and windows. Only remnants of paint are left on the northern and western wall of the baptistery. This condition is thought to be due to earlier water leakage and direct light exposure.

Despite the description above of damages in the painted surfaces, the paintings in the nave and baptistery of Kvernes are in remarkably good condition. One exception is a small area on the central part of the north wall in the nave where the décor was repainted. In this same area, where the paint layer is thicker, the paint was flaking before treatment by NIKU in 2013. Disintegration of the binding media was the main reason for carrying out the 2013 conservation treatments at Kvernes. Selected areas of the walls in the nave and baptistery were consolidated with 3% sturgeon glue in water applied through Japanese tissue paper. The aim of the treatment was to add an additional binder to the paint structure. In areas with two paint layers, it was also necessary to bind the layers to each other and to the wall. Unlike treatments in several other stave churches, the distemper decorative paintings in the nave and baptistery were not consolidated during an earlier conservation campaign. The consolidation of the distemper paintings in Kvernes stave church seems to be successful, except for minor areas where the paint layer is thick. Based on experience, this is not unexpected. In general, thick distemper paint layers are more prone to flaking.

Distemper paintings in unheated stave churches, like Kvernes, are generally in better condition than those in heated stave churches. NIKU's records show that the distemper painting (ca. 1601) found in the chancel of the unheated Urnes stave church (see Figure 4c for the Urnes site location) was consolidated in 2012, four centuries after it was painted. Then the painted ceiling in the chancel needed a general consolidation, while only minor areas of the painted chancel wall were treated in 2012. The unheated Rødven stave church (latitude: 62.62° N; longitude: 7.49° E) is in the same area of Norway as Kvernes. The interior is covered with distemper paintings from the first half of the 18th century which are worn, but in a rather good state, and have never undergone consolidation treatment.

2.2. Effects of Climate-Induced Decay on Distemper Paints

In general, the current conservation state of distempered 17th century paintings in unheated churches is better than those found in heated churches. When comparing the conservation state of Medieval art in heated and unheated churches this also holds true [20]. This demonstrates the effect of a 'Proofed fluctuation' that, over the centuries, is caused by natural slow climate change and normal building use with passive systems [21]. However, the findings in this paper also demonstrate that the risk of physical damage, beyond that already accumulated from T and RH fluctuations, may be caused by the introduction of new variables such as heating systems and increase in visitors. The use of heating systems or high numbers of visitors may exceed the already experienced climate pattern, thus introducing new risks.

Today, the number of visitors to the remote Kvernes stave church is high. One might expect both for Kvernes and the other stave churches that the number of visitors and arrangements will continue to rise, causing a continuous change in the use of these outstanding and vulnerable buildings. Several of the stave churches are museum churches, housing special events and open for tourists in the summer season. In the case of Kvernes, today, there are around 13,000 visitors per season. Groups of up to 90 people that visit Kvernes are placed in the pews by guides; they may lean towards the distemper painted walls and contribute to the degradation of the paint by intentionally or unintentionally touching and rubbing the painted surfaces. Visitors are also contributing to modifying water vapor concentration and particle matter concentration inside the church. These factors must be taken into consideration when dealing with hygroscopic materials, such as animal glue. These are ongoing factors that demand for new research. However, visitors are wanted, for what is a cultural heritage site without the connection to people, past and present? In addition, the revenue from these visitors is an essential part in maintaining and conserving these sites. It is the correct balance between a church left alone, and a visited church that adds both to conservation of the church and to a meaningful experience for visitors.

Looking at the causes of good conservation conditions in Kvernes, low temperature (i.e., ranging between 5 °C and 10 °C) may be one of these as it slows water vapor diffusion and hence increases the time of the wood response to RH variations [22]. This effect could account for the frequent observation that low-temperature storage of wooden works of art, for example, in unheated historic buildings, favors their good preservation [22]. On the other hand, the impact of heating on distemper decorative paintings is particularly evident in heated churches as is the case of the Ringebu stave church (latitude: 61.5° N; longitude: 10.17° E), a parish church in ordinary use, situated in a dry inland climate, with sporadic heating when in use. When a conservation treatment was undertaken in 2010, monitoring of the internal RH of the church showed extraordinarily low values during heating episodes in the winter. Unlike Kvernes, the distemper decorative paintings in Ringebu are 1921 replicas of 18th century distemper decorative paintings. Specifically, these 1920s replicas are painted on top of a whitewash which covers the remnants of the original 18th century paint. In 2010, a rather thick layer of paint was consolidated with sturgeon glue, however, a few years later flaking paint was observed and continues to flake to this day. The paint layer is influenced by the fluctuating RH and T and the assumed different dimensional changes of the individual layers in the structure induce stresses, which cause cracking and flaking of the ground and paint layers [22]. In Ringebu the paint flakes between the various layers of paint.

Effects of Climate-Induced Decay on Distemper Paints. Is the Consolidation Changing the Paints Response to Fluctuations in Relative Humidity and Temperature?

Sturgeon glue, as a proteinaceous material, will most probably react to variations in RH and T in the same way as the original glue in distemper paint. Animal glue experiences a considerable dimensional change with a change in moisture content. Some animal glues can swell as much as 6% over the 0%–90% RH range [22]. Analysis of the properties of sound rabbit-skin glue films showed that they can withstand fluctuations of ±15% RH at 50% RH, but only ±8% at 35% RH. Like many organic materials, the moisture response of glue is relatively flat at moderate RH values (40%–60%), reducing both its response to RH fluctuations and the resulting stresses [23] (p. 34).

The added glue, from consolidation, will change the ratio of pigment to glue expressed as the pigment-volume concentration (PVC). PVC is defined as the ratio of pigment volume and total paint volume. Higher PVC means less glue, which means lower dimensional response to RH variations. PVC values of distemper paint found in Norwegian stave churches are unknown, but a matte paint generally implies a high PVC. A 75% pigment-volume concentration gives a matte paint [13]. The consolidated paint is still as matte as it was before treatment; however, it is assumed that the PVC of consolidated paint is lower since additional glue is added. The added sturgeon glue might lead to larger dimensional change, as it is hygroscopic and reacts like animal glue to climatic fluctuations. That said, NIKU has not reported that consolidation with sturgeon glue of a thin, matte distemper paint accelerated its degradation.

2.3. Standardized Risk Assessment Method Applied to Climate-Induced Risk on Distemper Paints

When evaluating a management plan for an indoor climate, in general, it is difficult to decide when to adopt a change and how to identify which modification must be implemented for improving the conservation conditions of a cultural heritage site or object. For stave churches, these types of discussions are left in the hands of heritage managers and stave churches managers. These stakeholders might be highly attracted to finding a balance between the optimum conditions for conservation, the available budget, and the requests of acceptable comfort by churchgoers or visitors, as well as what measures are technically and aesthetically possible when it comes to the church and interior. However, it could be beneficial for these stakeholders to visually understand how the risk of decay (single and/or synergistic) appears and evolves over the years. This could be helpful in taking action, for example, asking for the support of experts in proposing a modification in the environmental management of the church. Yet risk indices may be difficult to track without tools, techniques, documentation, and information systems. A risk assessment tool (RAT) helps to bridge this gap. RAT is a general term which includes tools for risk management which allow uncertainty to be addressed by identifying and generating qualitative or semi-quantitative metrics, prioritizing, and developing practices to track risks with greater probability. The modified RAT of Anaf et al. [7] was applied in this study and emphasizes the influence of temperature and relative humidity. This RAT is beneficial for conservators, allowing them to monitor ongoing decay processes of distemper decorative paints under different scenarios (e.g., passive and active climate control strategy, high and low visitor impact). Measures of risk management are more easily taken in an already heated church.

The correlations between degradation and extreme RH and T values (as in our proposed case study for Kvernes), and large RH fluctuations are described in literature for specific types of materials and heritage objects [22,24,25]. Specifically, for heritage objects, the proposed RAT uses a best-fit dose-response, or mathematical functions, to prioritize the agent of deterioration (e.g., through the weighted average approach; Figure 6) and define the damage thresholds which allow the assessment of risk. In the presented study, only T- and RH-induced risks on heritage objects resembling distemper decorative paint on wooden substrates were considered.

Figure 6 displays in the ordinate the weight per risk caused by risk agents (i.e., relative humidity (%) and temperature (°C)), which cannot be too low, too high, or with excessive fluctuations without increasing the risk of damage. In this study, Figure 6 represents the RAT to estimate three different categories of risk: biological decay (top plots), mechanical and chemical decay (bottom, left plot), and mechanical decay only (bottom, right plot). These risks are estimated following the approach introduced in a study by Anaf and co-workers [7] by comparing the data logger measurements recorded in Kvernes, with the corresponding risky target values/thresholds available in the literature [22,24,25]. Risk thresholds of unsafe RH fluctuations (bottom-right plot, external areas to the safe band) are obtained by the standard EN15757:2010 [26]. The SyMBoL project, with the implementation of two long-term monitoring campaigns in Heddal (latitude: 59.57° N; longitude: 9.17° E) and Ringebu stave churches (i.e., longer than two years), will soon provide data to better describe the "historic climate" of other stave churches under different protocols for heating use [26].

The RAT works as follows: in Figure 6, the climate-induced risk is quantified by converting the corresponding measured values of T and RH into a level of risk that is evaluated in a predefined scale of 0 to 1. That is to say that this model displays the correlation between X and Y, where Y = f(X). When T or RH x-axis values (input) enter the plots in a white area over the abscissa, the level of (risk*weight) is zero (output, y-axis); while, when T and RH x-axis values enter the plots within a grey area over the abscissa, the level of (risk*weight) is different from zero. The (risk*weight) value, read over the y-axis, is obtained moving vertically from the input T or RH value over the abscissa up to the parallel grey horizontal line limited by two grey dots in Figure 6.

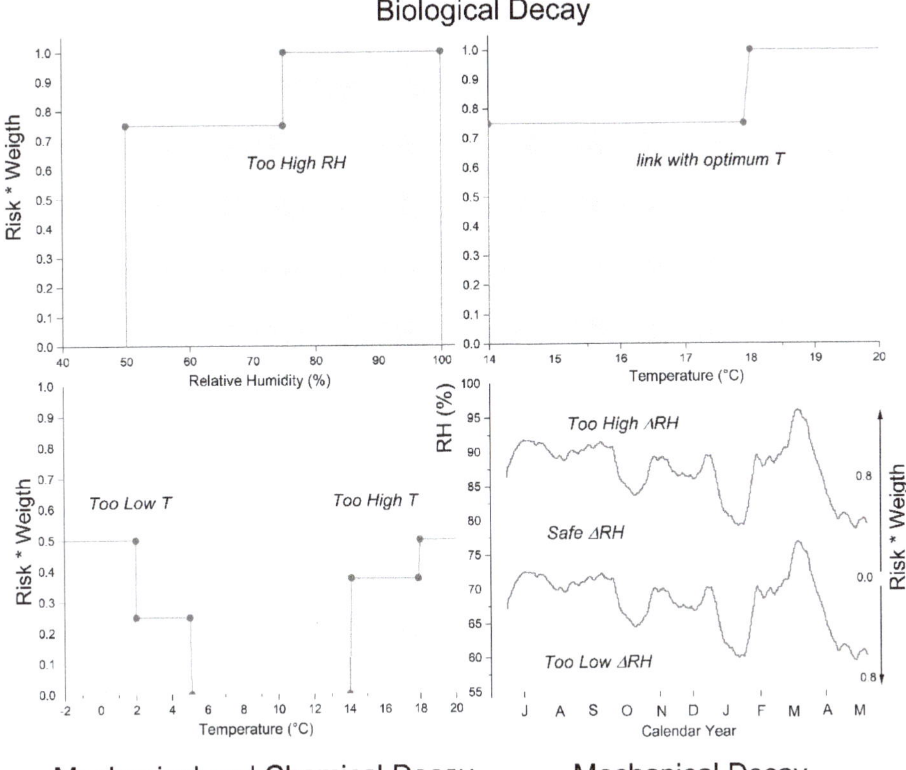

Figure 6. Conversion dose-response and/or threshold functions to prioritize the agent of decay and calculate the level of risk. (top-left) relative humidity (RH)-induced biological decay in the case of both RH-sensitive insects as woodworm and mold infestation with their link with temperature (T) conditions (top-right). Bottom-left: T-induced mechanical (too low T) and chemical risk of decay (too high T). Bottom-right: RH-induced mechanical decay caused by wider RH fluctuation than the historic climate as stated in the EN15757:2010 standard [26].

In Kvernes church, the T and RH values were recorded using four Tinytag data loggers (Gemini Data Loggers Ltd., Chichester, UK) deployed on the east side of the church (Figure 7). The positioning of these four loggers was slightly different; data loggers 1 and 2 were placed nominally 2 m above the floor, while data loggers 4 and 5 were placed about 3 m above the floor. Temperature and relative humidity data were recorded for a little over one year (20/06/2011–13/06/2012), after which, the data collected during this monitoring campaign were analyzed using the RAT.

The abscissa of the grey dots are T and RH target/threshold values which define unsafe conditions within which the (risk*weight) is different from zero.

Figure 7. Images showing the location of the four data loggers in Kvernes stave church. (**a**) loggers 1 and 2; (**b**) loggers 4 and 5 (date: 3 March 2011) Photo: NIKU © 2011.

3. Results and Discussion

Temperature and relative humidity data collected during 20/6/2011–13/6/2012, from the interior of Kvernes church, are shown in Figures 8 and 9, respectively. The right *y*-axis reports the risk assessment results of T-induced risk (Figure 8) and RH-induced risk (Figure 9). For this study, the maximum risk value was pre-set to 1 for T-induced risk (right *y*-axis of Figure 8) and a value of 1.5 (over a scale of 2.0) for RH-induced risk as we counted both biological (max value = 1) and mechanical risk (max value = 1) (right *y*-axis of Figure 9). The grey lines in Figures 8 and 9 denote the level of risk, and Figure 8 clearly indicates that the riskiest period for temperature-driven decay is during the winter/spring period. Additionally, data acquired by Log5 indicate greater variations, especially in the spring/summer seasons in term of maxima registered temperatures, since this logger was exposed to direct sunlight. The positioning of Log5 was deliberate, in order to assess the effect of solar radiation on the sensors. Consequently, the relative humidity data recorded by Log5 showed the highest variations, and constantly registered lower values with respect to the three other devices. On the contrary, data collected from Log1 often indicated oversaturation of RH values (>100% RH) (Figure 9). The risk analysis carried out on the RH data (Figure 9) is more complex, as it displays intermitted climatic conditions during autumn, between mid-September and the beginning of December. However, most of this time period could be considered quite safe (risk = 0).

Unfortunately, very few studies are currently available in international peer-reviewed literature to correlate the effect of microclimate variations to the occurrence of decay on distemper paints. The very few existing data on indoor microclimatic conditions of wooden Medieval churches seem to differ [27,28]. One possible answer to the variation in the indoor climate is the outdoor climate where the various stave churches are located (see the Supplementary Materials for our analysis of the outdoor climate in Kvernes). One example is the average environmental conditions published very recently on the Hopperstad stave church [29], where RH ranged between 27%–28% and 92%–98%, depending on the placement of the data logger. Hopperstad is an unheated church situated on a fjord in the west of Norway (latitude: 61.01° N; longitude: 6.95° E) and in a similar climatic zone as Kvernes [30]. Maximum and minimum temperature values for Hopperstad, indicated by Lehne et al. in 2019 [29], seem closer to those collected for Kvernes stave church in this study. The influence of solar radiation on the response of the data logger, in our study, is not too extreme as Figure 8 clearly demonstrates because the internal temperature never exceeds 22 °C and never falls below −2.5 °C. On the other hand, RH reading reached saturation values (>100% RH) several times between mid-December 2011 and the end of March 2012 (Figure 9). While at the opposite, the minimum RH values never fell below 42.5%. Hopperstad has remnants of distemper decorative paint which was consolidated by NIKU in 2010.

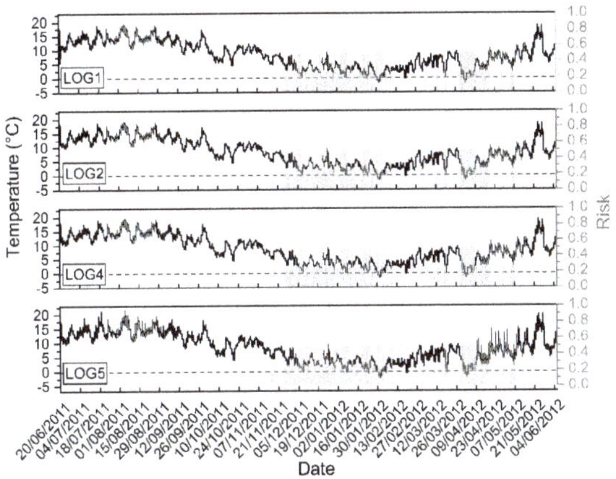

Figure 8. Indoor temperature data from the four data loggers collected between 20/6/2011 and 13/6/2012 every 2 h. Dotted lines indicate T = 0 °C. Grey vertical lines indicate the level of risk (right *y*-axis).

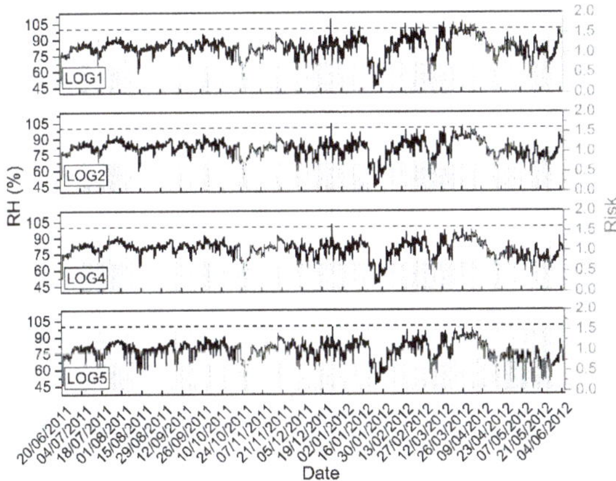

Figure 9. Indoor relative humidity (RH, %) data from the four data loggers collected between 20/6/2011 and 13/6/2012 every 2 h. Dotted lines indicate RH = 100%. Grey vertical lines indicate the level of risk (right *y*-axis).

When examining data collected from Log1, Log2, and Log4, we decided to use the data from Log4 as the representative of this group, as the data collected from these three loggers are directly comparable (Figures 8 and 9); consequently, as Log5 was constantly hit by sun radiation, it was considered separately. Data from Log4 and Log5 were elaborated using a RAT (Figures 10 and 11). Figure 10 highlights the combined effect (Figure 10d) of the indoor environmental conditions and the risk threshold of the three most important agents of deterioration for distempered paint: biological decay (Figure 10a), mechanical decay due to the RH fluctuation (Figure 10b), and mechanical decay induced by low T values (Figure 10c). One might argue that the impact of RH fluctuations outside the frame of the historic climate should be placed as the primary deterioration agent. However, in cases like Kvernes, it is difficult to decide which agent of deterioration is the most important. During the monitored calendar year, starting in June of 2011, risk of biological decay was recorded in the summer months, with a maximum at the end of July (Figure 10a). Risk of mechanical decay caused by a RH variation, larger than that established by the historic climate, was recorded from January to March (Figure 10b). In addition, risk of mechanical decay caused by too low T (<2 °C), was slightly higher from December to February (Figure 10c). Overall, the synergistic risk recorded in Kvernes highlights three periods of risk (Figure 10d). The first risk period occurs during July and August showing the optimum condition for biological decay due to high T and RH values; whereas, the second risk period starts at the beginning of January and for almost a whole month shows the optimum conditions for mechanical decay on the distemper paint due to a combination of large RH fluctuations and very low temperature (i.e., lower than 2 °C). The third and final risk period occurs in May, and this period highlights a risk of mechanical decay caused by RH fluctuations only. At the opposite, the period with natural environmental conditions most appropriate for the conservation of the distempered paint happens in the autumn, from September to November, and spring in mid-April. Other aspects, like the actual T of the wall surface, needs to be considered before deciding upon a conservation campaign in November.

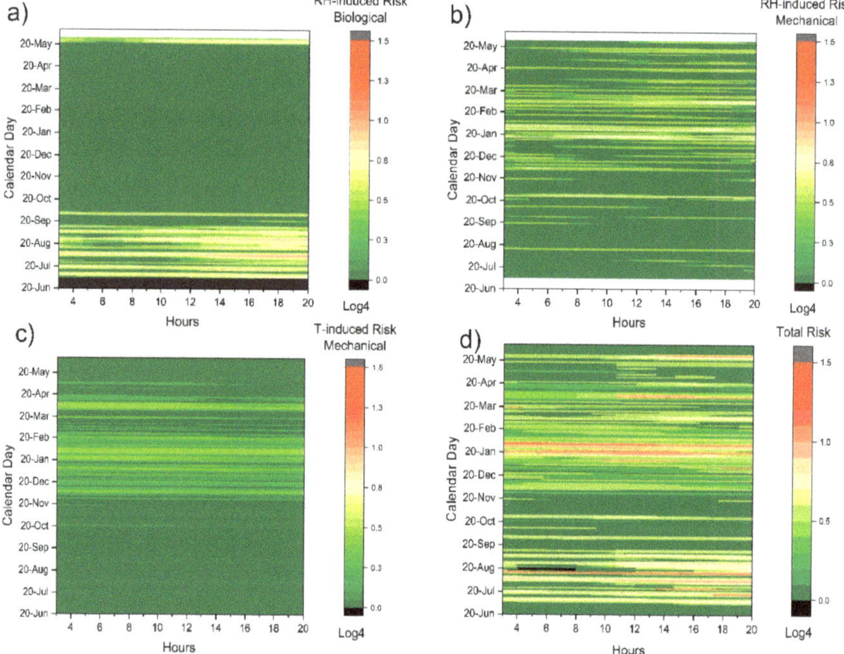

Figure 10. (**a**) RH-induced risk of biological decay on wood (i.e., woodworm and fungi infestation). (**b**) RH-induced risk of mechanical decay caused by wider RH fluctuation than the historic climate as stated in the EN15757:2010 standard; (**c**) T-induced risk of mechanical decay on wood (i.e., freezing–thawing cycles; (**d**) total risk (i.e., sum of the previous three types of calculated risk to show the synergetic effect. Colors associated to the maps are related to the index value reported on the scale: green from 0 to 0.5 indicates low risk conditions; yellow to orange, from 0.5 to 0.75 indicates moderate risk; red, approaching 1.5 indicates high risk.

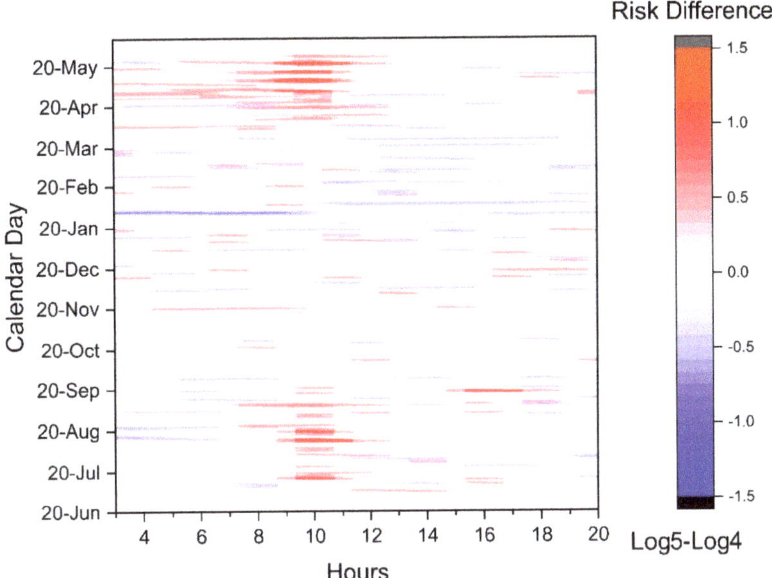

Figure 11. Risk difference caused by radiation disturbance as detected by the difference between Log5 and Log4. Red color, environmental conditions which trigger higher total risk; blue color, environmental conditions which trigger lower total risk.

The elaboration shown in Figure 11 was obtained by comparing the evaluated risks for Log5 and Log4, as discussed above. In this figure, the red areas denote an increase in risk, whereas the blue areas signify a decrease in risk. When examining only the red areas, it is possible to observe how the most sensitive periods occur during the summer months (mid-June to the beginning of September) and spring months (mid-March to mid-May). Additionally, Figure 11 also displays how the risky daily interval partially overlaps for both periods. During the summer the daily risk intervals are concentrated between 9:00 and 11:00 a.m., while in spring longer intervals are observed (from 2:00 to 11:00 a.m.). For just a short number of spring days (ca. 15) the whole morning period is considered risky. This risk assessment once stopped dangerous wavelength radiation, as UV and IR also provide interesting information on the beneficial effect of a daylight increase in temperature (risk decrease in blue areas).

The previous figures demonstrate that such a tool becomes a semi-quantitative method to recognize the effect of climate change (e.g., through the evaluation over time of changes in risk intensity, in frequency of less favorable conservative conditions, or in increasing synergistic effects). The future research conducted within the SyMBoL project will contribute to developing a more comprehensive RAT for conservation and preservation purposes. The SyMBoL project intends to integrate the impact of visitors, in terms of modification of the indoor microclimate, into the tool. In fact, RAT has the potential to highlight the effect of visitors' presence (or the effect of heating systems. These are important aspects, which although not considered in the present study, must be considered in the perspective of long-term research of stave church conservation.

Finally, in order to develop a broader understanding of the internal microclimate of Kvernes, a detailed analysis was performed for Log4 and Log5 on the weeks of Christmas and Easter when the Church hosts churchgoers. Elaborations proposed in Figures 12 and 13 show windows segmenting the daily light durations as indicated by data available at eKlima website of the Norwegian Meteorological Institute [31], superimposed to T, RH, and mixing ration (MR) trends (i.e., the trend of water vapor (gr) mixed into 1 kg of dry air which is a useful microclimate parameter used in historic structures to study the effects of infiltration).

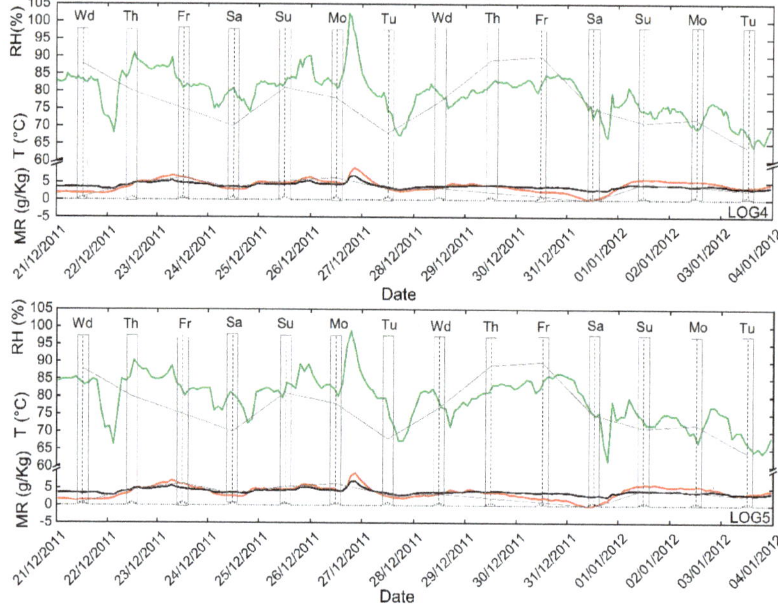

Figure 12. Indoor RH (%)-green line, T (°C)-red line, mixing ration (MR) (g/Kg)-black line values for Christmas week representative of microclimate conditions and churchgoers' presence in winter as measured by Log4 (top plot) and Log5 (bottom plot). The grey line refers to outdoor RH values collected by the Kristiansund lufthavn weather station. Daylight duration is highlighted by rectangular areas.

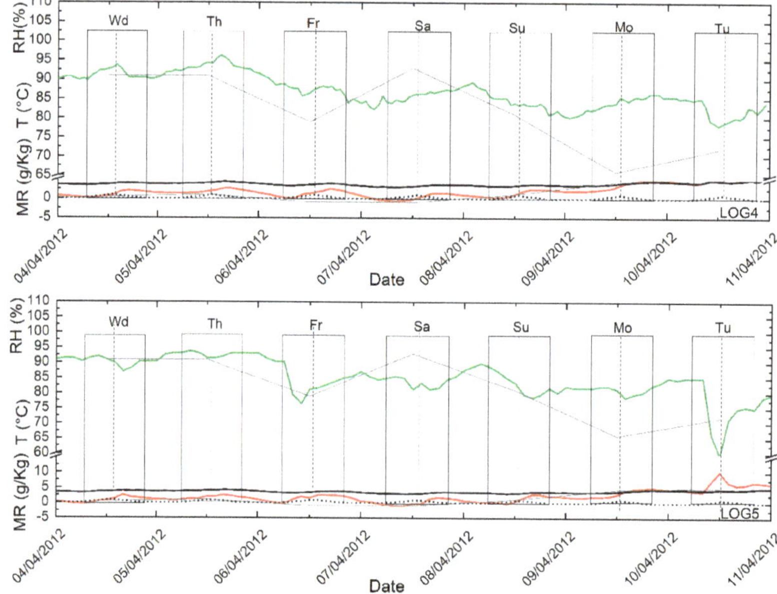

Figure 13. Indoor RH (%)-green line, T (°C)-red line, MR (g/Kg)-black line values for Easter week representative of microclimate conditions and churchgoers' presence in spring as measured by Log4 (top plot) and Log5 (bottom plot). The grey line refers to outdoor RH values collected by the Kristiansund lufthavn weather station. Daylight duration is highlighted by rectangular areas.

4. Conclusions

NIKU has treated distemper decorative paintings in 14 of the 28 stave churches, as well as in several other churches. The challenge of consolidating matte, water-soluble paint is finding a consolidation substance that binds loose paint and strengthens the paint layer without saturating the structure in such a way that it changes the look of the artwork. The consolidation agent must be compatible with the binder in the paint; an ideal consolidant should have known ageing characteristics and preferably decompose naturally, thus making reconsolidation possible. First and foremost, an appropriate consolidant must conserve the paint layer. Since the early 1990s sturgeon glue has been the dominant consolidation medium because it changes the visual appearance of the distemper paintings the least; it is a natural adhesive substance which decomposes in the same way as the original binder and has strong penetrative powers and a high degree of adhesion at low concentrations.

The applied consolidation method with sturgeon glue works well for thin paint layers which need additional binding medium. The conservation of thicker distemper paint, for example where there are two layers of decorative paint on top of one another, is still a challenge and NIKU is still searching for a method that gives a good visual result, whilst adhering the paint to the substrate. The consolidation of thick paint layers is a problem in both unheated and heated churches, but experience shows that the thick paint flakes faster after treatment in heated churches. Recent surveys of the paintings in the nave and baptistery show that on the north wall there might be some locations with paint lose in areas with two paint layers, while the condition for the rest of the paintings in the nave and baptistery appears unchanged.

The RAT proposed by Anaf et al. in 2019, which was modified for distempered paints, demonstrated that the indoor climate of Kvernes—unperturbed by any heating system and still remaining in its so-called natural state-although not completely ideal, contributes to the preservation of distemper paints. Unsurprisingly, the indoor climate follows the outdoor climate without any control on keeping microclimate targets and/or reducing fluctuations and showed a limited number of risk periods for Kvernes over 2011–2012. This assessment, although not exhaustive of the impact of visitors or particle matter deposition phenomena, provides the potentiality to highlight at once the most and least favorable conservation conditions for distemper decorative paint over a typical calendar year. This type of understanding supports heritage managers and church owners in identifying risks situations and discriminating among the possible typologies of decay. In addition, the output of this analysis provides reliable information for supporting a risk management plan for the preservation of distemper decorative paint. The SyMBoL project coordinated by NTNU is working for both the sake of cultural heritage and for those who want to experience the wooden construction and vivid colors.

Supplementary Materials: The following SM are available online at http://www.mdpi.com/2225-1154/8/2/33/s1.

Author Contributions: T.M.O.: writing part in Sections 1.1–1.3, except for the first introduction, contributed to the editing of the text, contributed climate logger registrations in Kvernes stave church, contributed to Section 2, as well as discussion and conclusion sections. A.A.Ø.: contributed to Section 2, reference list, and responsible for figures. N.K.J.: contributed to the writing of Sections 1 and 2, contributed to reference list. L.d.F.: analysis and elaboration of microclimatic data from installed loggers, part of introduction section, result section on data loggers, and contributed to text formatting and editing. A.F.: writing of the abstract, and contributed text formatting, editing, and revisions. C.B.: paper coordination, risk assessment data elaboration and plots creation, writing of results and discussion and conclusions sections, contributed in the revision. All authors have read and agreed to the published version of the manuscript.

Funding: This study was funded by the Norwegian Research Council within the framework of the "SyMBoL— Sustainable Management of Heritage Building in a Long-term Perspective" Project (Project No. 274749).

Acknowledgments: The authors are grateful to Else Marie Bae of the Fortidminneforeiningen for her kind support during the monitoring campaign and to the Norwegian Directorate for Cultural Heritage.

Conflicts of Interest: The authors declare no conflicts of interest.

References

1. Varas-Muriel, M.J.; Martínez-Garrido, M.I.; Fort, R. Monitoring the thermal–hygrometric conditions induced by traditional heating systems in a historic Spanish church (12th–16th C). *Energy Build.* **2014**, *75*, 119–132. [CrossRef]
2. Camuffo, D.; Pagan, E.; Rissanen, S.; Bratasz, L.; Kozlowski, R.; Camuffo, M.; della Valle, A. An advanced church heating system favourable to artworks: A contribution to European standardization. *J. Cult. Herit.* **2010**, *11*, 205–219. [CrossRef]
3. Napp, M.; Wessberg, M.; Kalamees, T.; Brostrom, T. Adaptive ventilation for climate control in a medieval church in cold climate. *Int. J. Vent.* **2016**, *15*, 1–14. [CrossRef]
4. Larsen, P.K. Climate control in Danish churches. In *Proceedings of Museum Microclimates*; Padfield, T., Borchersen, K., Eds.; National Museum of Denmark: Copenhagen, Denmark, 2007.
5. Ørstavik, R. *Ørsta Kyrkje Gjennom Tidene*; Ørsta Kulturstyre: Ørsta, Norway, 1990.
6. Haugen, A.; Bertolin, C.; Leijonhufvud, G.; Olstad, T.; Broström, T. A Methodology for Long-Term Monitoring of Climate Change Impacts on Historic Buildings. *Geosciences* **2018**, *8*, 370. [CrossRef]
7. Anaf, W.; Pernia, D.L.; Schalm, O. Standardized Indoor Air Quality Assessments as a Tool to Prepare Herit-age Guardians for Changing Preservation Conditions due to Climate Change. *Geosciences* **2018**, *8*, 276–290. [CrossRef]
8. Olstad, T.M. To the Glory of God and the Church's adornment. In *Preserving the Stave Churches*; Bakken, K., Ed.; Pax Forlag: Oslo, Norway, 2016.
9. Finance Norway 2019: Climate Report Finance Norway 2019. Available online: https://www.finansnorge.no/siteassets/tema/barekraft/klimarapport-finans-norge-2019.pdf (accessed on 8 January 2020).
10. Olstad, T.M.; Berg, F. Hvorledes Sikre Og Forvalte Norske Kirkebygninger i Fremtidens Klima. A 335 Kvernes Stavkirke: 19. NIKU Oppdragsrapport 150/2016. Available online: http://hdl.handle.net/11250/2583623 (accessed on 8 January 2020).
11. Borgen, L.W. Har Stavkirkene mer å Fortelle? Fortidsminneforeningens Jubileumsseminar.16.des. 2019. Oral Presentation. The Neighboring Grip Stave Church is Probably Dated 1620s, See also Storsletten, O.; Anno 1621 Bleff Denne Kircke Bygd. En undersøkelse av Grip stavkirke, NIKU Rapport Bygninger og Omgivelser 15/2007 NIKU 2007. 2019. Available online: https://ra.brage.unit.no/ra-xmlui/bitstream/handle/11250/176379/Stavkirkeprogrammet_Grip_NIKUOppdragsrapport_128_2009.pdf?sequence=1&isAllowed=y (accessed on 13 February 2020).
12. Olstad, T.M.; Solberg, K. Eight seventeenth -century decorative paintings- one painter? Studies in Conservation. In *Painting Techniques History, Materials and Studio Practice: Contributions to the Dublin Congress*; Ashok, R., Smith, P., Eds.; International Institute for Conservation of Historic and Artistic Works: London, UK, 1998.
13. Hansen, E.F.; Walston, S.; Bishop, M.H. Matte Paint. In *Its History and Technology, Analysis, Properties, and Treatment, with Special Emphasis on Ethnographic Objects*, 1st ed.; Oxford University Press: Oxford, UK, 1994.
14. Brænne, J.; Havrevold, M. *Befaringsrapporter Fra Stavkirker*; The Directorate for Cultural Heritage, Antiquarian Archives, Unpublished report, 1984.
15. Olstad, T.M.; Ørnhøi, A.A. *Konsolidert Limfargedekor i Stavkirkene-en Oversikt. Vurdering av Størlimkonsolidering av Limfarge på tre. Delprosjekt 1*; NIKU Oppdragsrapport 62/2014; NIKU: Oslo, Norway, 2014. Available online: https://docplayer.me/16381408-Konsolidert-limfargedekor-i-stavkirkene-en-oversikt.html (accessed on 13 February 2020).
16. Solberg, K.; Olstad, T. Konsolidering av limfargedekor. In *Surface Treatments: Cleaning, Stabilization and Coatings, Proceedings of IIC Nordic Group, Danish Section XIII Congress*; National Museum of Denmark: Copenhagen, Denmark, 1994.
17. Brænne, J. *The Synagogue from Horb in Germany. Examinations and Suggestions for Treatment. Jerusa-lem/Oslo*, Unpublished report, 1987.
18. Olstad, T.M.; Solberg, K. Limfargedekor på tre-analyser og observasjoner. In *Strategisk Instituttprogram 1996–2001. Konservering: Strategi og Metodeutvikling*; Swensen, G., Ed.; NIKU Publikasjoner: Oslo, Norway, 2001.
19. Olstad, T.M.; Kaun, S. *Konserveringsarbeid i Kvernes Stavkirke Sommeren 2012. A 335 Kvernes Stavkirke, Averøya, Averøy Kommune*; NIKU Oppdragsrapport 192/2012; NIKU: Oslo, Norway, 2012; Available online: https://docplayer.me/65543819-Kvernes-stavkirke-avsluttende-arbeider-i-interioret.html (accessed on 13 February 2020).

20. Olstad, T.; Stein, M. *Saving Art by Saving Energy*; NIKU-Norwegian Institute for Cultural Heritage Research: Oslo, Norway, 1996.

21. Michalski, S. The Ideal Climate, Risk Management, the ASHRAE Chapter, Proofed Fluctuations, and Toward a Full Risk Analysis Model. In Proceedings of the Contribution to the Experts' Roundtable on Sustainable Climate Management Strategies, Tenerife, Spain, 14 April 2007.

22. Bratasz, L. Allowable microclimatic variations for painted wood. *Stud. Conserv.* **2013**, *58*, 65–79. [CrossRef]

23. Erhardt, D.; Mecklenburg, M. Relative humidity re-examined. In *Preventive Conservation. Practice, Theory and Research, Preprints of the Contributions to the Ottawa Congress*; Roy, A., Smith, P., Eds.; IIC: London, UK, 1994.

24. Mecklenburg, M.F.; Tumosa, C.S. Mechanical behavior of paintings subjected to changes in temperature and relative humidity. Studies in the Transport of Paintings. In *International Conference on the packing and Transportation of Paintings*; Art in Transit: Washington, DC, USA, 1991.

25. PAS 198:2011 Specification for Environmental Conditions for Cultural Collections. Available online: https://icon.org.uk/system/files/documents/pas198_icon_ag_response_1.pdf (accessed on 13 February 2020).

26. EN 15757: 2010. *Conservation of Cultural Property–Specifications for Temperature and Relative Humidity to Limit Climate-Induced Mechanical Damage in Organic Hygroscopic Materials*; European Committee for Standardization: Brussels, Belgium, 2010.

27. Marstein, N.; Stein, M. Advanced Measuring of the Climatic Conditions in the Medieval Wooden Churches in Norway. In Proceedings of the ICOM Committee for Conservation: 8th Triennial Meeting, Sydney, Australia, 6–11 September 1987.

28. Olstad, T.M. Medieval Wooden Churches in Cold Climate-Parish Churches or Museums? In *Preventive Conservation: Practice, Theory and Research, Preprints of the Contributions to the Ottawa Congress*; Roy, A., Smith, P., Eds.; IIC: London, UK, 1994.

29. Lehne, M.; Mantellato, S.; Aguilar Sanchez, A.M.; Caruso, F. Conservation issues and chemical study of the causes of alteration of a part of the Stave Church in Hopperstad (Norway). *Herit. Sci.* **2019**, *7*, 80–98. [CrossRef]

30. Climate Data Norway. Available online: http://www.senorge.no/index.html?p=klima (accessed on 8 January 2020).

31. Available online: http://sharki.oslo.dnmi.no/portal (accessed on 3 February 2020).

Article

Investigation on the Use of Passive Microclimate Frames in View of the Climate Change Scenario

Elena Verticchio [1,*], Francesca Frasca [2] and Fernando-Juan Garcìa-Diego [3] and Anna Maria Siani [2]

[1] Department of Earth Sciences, Sapienza Università di Roma, P.le A. Moro 5, 00185 Rome, Italy
[2] Department of Physics, Sapienza Università di Roma, P.le A. Moro 5, 00185 Rome, Italy
[3] Department of Applied Physics, Centro de Investigación Acuicultura y Medio Ambiente ACUMA, Universitat Politècnica de València, 46022 Valencia, Spain
* Correspondence: elena.verticchio@uniroma1.it; Tel.: +39-06-4991-3479

Received: 28 June 2019; Accepted: 5 August 2019; Published: 9 August 2019

Abstract: Passive microclimate frames are exhibition enclosures able to modify their internal climate in order to comply with paintings' conservation needs. Due to a growing concern about the effects of climate change, future policies in conservation must move towards affordable and sustainable preservation strategies. This study investigated the hygrothermal conditions monitored within a microclimate frame hosting a portrait on cardboard with the aim of discussing its use in view of the climate expected indoors in the period 2041–2070. Its effectiveness in terms of the ASHRAE classification and of the Lifetime Multiplier for chemical deterioration of paper was assessed comparing temperature and relative humidity values simultaneously measured inside the microclimate frame and in its surrounding environment, first in the Pio V Museum and later in a residential building, both located in the area of Valencia (Spain). Moreover, heat and moisture transfer functions were used to derive projections over the future indoor hygrothermal conditions in response to the ENSEMBLES-A1B outdoor scenario. The adoption of microclimate frames proved to be an effective preventive conservation action in current and future conditions but it may not be sufficient to fully avoid the chemical degradation risk without an additional control over temperature.

Keywords: microclimate frame; preventive conservation; risk assessment; Sorolla painting; climate change

---

1. Introduction

The environment surrounding the objects is one the main driver of their deterioration. Long-term microclimate monitorings, through the identification of risk factors, play a key role in the implementation of preventive conservation actions [1]. Temperature and relative humidity are fundamental physical parameters, as materials adapt themselves to the continually changing hygrothermal conditions to reach a thermodynamic equilibrium. Strict microclimate targets for preservation [2,3] have fostered the use of expensive HVAC (Heating, Ventilation and Air-Conditioning) systems. However, these highly sophisticated systems may be risky in the case of a potential failure and hardly possible for all museums, which, with the increase of cultural tourism, might incur raised costs for the maintenance of adequate conditions for conservation [4]. Furthermore, due to a growing concern about the effects of the expected climate change, future policies in conservation must move towards affordable and sustainable preservation strategies [5]. Passive methods, based on the understanding of the material properties and of its interaction with the environment, might provide a reliable support in this direction.

Among preventive conservation tools, showcases aim at creating an internal micro-environment different from the external macro-environment [6]: this "box-in-box" configuration allows locally fine-tuning the control over various environmental parameters (temperature, relative humidity, pollutants and light), thus reducing the risk of physical and chemical damage to cultural heritage objects [7]. The employment and optimization of passive low-cost devices can be highly effective to provide relative humidity control in less than ideal environments, particularly in the case of mixed collections with different conservation needs. Since their response to temperature fluctuations is usually poor [6,8], panels of materials containing PCMs (Phase Change Materials) have been proposed to be placed inside showcases to keep the internal temperature stable [9].

Microclimate frames are showcases specifically designed for paintings and able to modify their internal conditions in order to comply with tolerability targets for specific typologies of materials. This kind of exhibition enclosures is considered among the safest systems for keeping relative humidity stable and is increasingly being used to protect paintings against indoor hazards [10]. Passive microclimate frames usually take advantage of the inclusion of a buffering agent in combination with the reduction of the air exchange rate [11–13]. A buffering agent is generally an extremely absorbent material which is able to smooth out abrupt changes by releasing moisture when relative humidity decreases and absorbing moisture if it increases. An economical microclimate frame can be produced in-house using the picture's frame as the primary case [14]. As in showcases, every microclimate frame is characterized by a peculiar response to the environmental forcing [8]. Their effectiveness depends on the specific features and can be assessed as a function of the improvement of the surrounding microclimate in terms of the fulfilment of the artwork conservation needs.

The most recent standards in conservation avoid recommending ideal temperature and relative humidity intervals and have evolved towards the concepts of proofed fluctuations [15], i.e., the largest hygrothermal levels experienced by the objects in the past, and historic fluctuations [16], i.e., the environmental conditions to which artworks have acclimatized and adapted during their conservation history. Both these concepts imply methodological indications rather than prescriptive ones [3] and thus a more flexible approach, allowing for the short-term fluctuations and seasonal changes that can be considered safe for the collections. The ASHRAE (American Society of Heating, Air-Conditioning and Refrigerating Engineers) guidelines [17] suggest five classes of quality control, defined on the basis of seasonal and daily hygrothermal fluctuations. The possible risks for collections gradually increase from Class AA, associated with no risk to most objects, to Class D, that protects only from dampness. These guidelines have been effectively applied to quantify the damage potential of environments already actively controlled [18] and those of future climate scenarios [19]. Thanks to the enhanced knowledge of the properties of the materials and of the mechanisms of interaction with the surrounding environment, damage functions can be used to assess the possible risks for various typologies of materials [18]. For paper, one of the most alarming degradation processes is the chemical decay (e.g., yellowing of paper and fading of colors) [20]. The Lifetime Multiplier is an index extensively used to assess the time span in which varnishes and paper objects remain usable if compared to standard reference conditions [21,22].

To extend the microclimate assessment over the effects of the expected climate change, simplified heat and moisture transfer equations through the building envelope can be derived from monitored outdoor and indoor data and employed to simulate the future conditions indoors [23,24]. This methodology was developed within the European project Climate for Culture (2009–2014) [25,26], which focused its attention to the future conservation risks with the aim to suggest possible mitigation actions and inform stakeholders and policy makers.

Simultaneous measurements of temperature and relative humidity collected inside and outside a microclimate frame were used in this study to investigate the quality of its internal environment, making it possible to evaluate the buffering properties over time. The hygrothermal observations were recorded from May 2014 to February 2017, first in the Sorolla room of the Pio V Museum of Fine Art in Valencia (Spain) [27] and, later on, in a residential building in the same area. The effectiveness of the passive microclimate frame was expressed in terms of the ASHRAE classification and the Lifetime Multiplier index for chemical deterioration. Moreover, a methodology based on heat and moisture transfer functions through the building envelope was applied to derive projections over the future indoor hygrothermal conditions as a function of the ENSEMBLES-A1B outdoor scenario in the area of Valencia [28]. An increased awareness of the potential conservation risks in view of the expected climate change has given the possibility to suggest appropriate preventive conservation strategies.

2. Materials and Methods

2.1. The Microclimate Frame

A long-term hygrothermal monitoring was conducted inside a microclimate frame housing a portrait of the Valencian painter Joaquín Sorolla (1863–1924). The painting, titled "Portrait of a lady with a red flower in her hair" (Figure 1a), measures 64 cm × 49 cm and is enclosed in a hand-crafted microclimate frame (69 cm × 54 cm ×8 cm) made of an external aluminium case and a frontal glass. The specific layout of the components of the microclimate frame under study is shown in Figure 1b. A sheet of cardboard (i.e., the same material supporting the portrait) of the same size of the paintings was used as back plate for the frame and put in direct contact with the painting support in order to offset changes in external relative humidity acting as a buffer [14]. The cardboard was preconditioned to the relative humidity level of 40% according to an extensive literature review on paper degradation [20] with the aim of reducing the impact of deterioration risk factors acting on the painting.

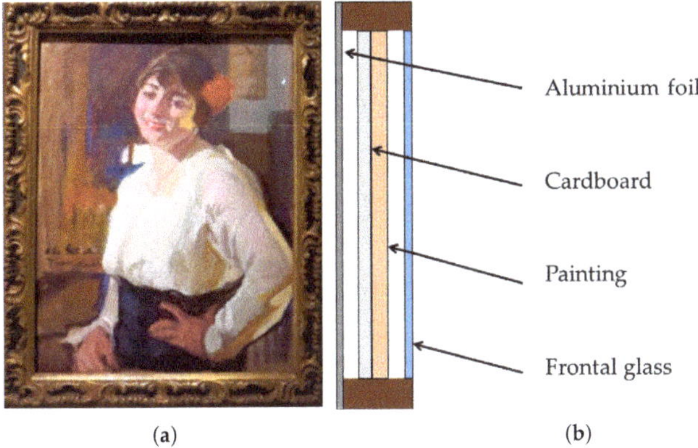

Figure 1. Portrait on cardboard titled "Portrait of a lady with a red flower in her hair" by J. Sorolla (**a**); and schematic cross section of the microclimate frame used and layout of its components (**b**).

The painting by Sorolla was realized on cardboard in 1916 with the gouache technique and donated as a gift to the Traver family. Since then, it used to be conserved in the house of the owners being enclosed within an unbuffered frame. Conservation surveys performed by the Valencian Institute of Conservation and Restoration (IVACOR) detected the presence of dust deposits both on the front and on the back, craquelures and loss of material on the painting layer together with a massive fungal attack visible in the form of dark circular stains. The gum arabic, frequently used as binding media [29], is responsible for its sensitivity to hygrothermal variations as it is particularly vulnerable to mold growth and chemical degradation. In the final report of the surveys, the Institute warned that the deterioration could have been caused by adverse environmental conditions in conjunction with the vulnerability of the materials used [30]. The artwork underwent restoration from 2012 to 2014 in the IVACOR laboratories and at the end of the intervention was enclosed in a passive microclimate frame *ad hoc* designed and provided with internal temperature and relative humidity sensors.

2.2. The Monitoring Campaign

Temperature (T) and relative humidity (RH) data were monitored from May 2014 to February 2017. Over this period, the painting was exposed to different environments: first, the Sorolla room in the Pio V Museum of Fine Art in Valencia from May 2014 to February 2016 and, later on, a residential building located in a city near Valencia. Since the private owners do not want to reveal the new location of the painting for safety reasons, in this investigation were used the climate data of the area of Valencia. The Pio V Museum is housed in a historical building of the XVII century where an active HVAC system of temperature control was in operation, with a variable T set-point ranging from 20 °C to 24 °C and RH left uncontrolled [27]. In the residential building, where the painting continued to be monitored with the same T and RH probes, an intermittent heating system was active only in winter and temperatures exceeded 30 °C during summer.

The microclimate monitoring system was developed by the Department of Applied Physics of the Polytechnic University of Valencia [31]. Two probes, each with coupled T and RH sensors, were assembled and installed within and outside the microclimate frame. Some of the technical features of the sensors are reported in Table 1: the temperature sensors (Maxim Integrated DS18B2) are in accordance with the instrumental metrological characteristics recommended in EN 15758:2010 [32], while the uncertainty of the RH sensors (Honeywell HIH 4030) is slightly higher than that recommended by EN 16242:2012 (3%) [33]. When using multiple sensors for RH, they must be carefully calibrated in advance in order to have no significant difference in their accuracy. For this reason, the RH sensors were calibrated with aqueous solutions of two salts (lithium chloride and sodium chloride) in accordance with the ASTME 104-02 standard [34]. The time interval between consecutive observations was set to 1 h, following the results of a previous study [27] where the sampling frequency was found to be reliable in the application of recent standards and therefore can be considered a good compromise between the priority of disposing of detailed series of observations and the necessity of avoiding redundancy in museum surveys.

The mixing ratio (MR) was derived from simultaneous T and RH data using the formula in [33].

Table 1. Technical features of the T and RH sensors used in the monitoring.

	T	RH
Response time	750 ms	5 s
Uncertainty	±0.5 °C	±3.5%

The outdoor hygrothermal data were obtained from the meteorological hourly dataset of the area of Valencia [35] distributed by the National Agency of Meteorology of the Spanish Government (AEMET) and available on the UPV website.

2.3. The Environmental Risk Assessment

The environmental risk assessment was based on the application of the ASHRAE guidelines and on the computation of the Lifetime Multiplier, an index used to quantify the risk of chemical degradation for paper.

In the ASHRAE guidelines [17], the classification in classes of quality of environmental control is based on the combination of the T and RH seasonal cycles and short-term fluctuations. The maximum and minimum seasonal shift is calculated by adding and subtracting to the annual mean the seasonal changes allowed for each class. The width of the final bands is finally determined shifting the curve of a 91-day central moving average by the short-term fluctuations indicated for the same class [18]. The ASHRAE classes of quality for conservation range from Class D, which prevents only from dampness, to Class AA, which is associated to no risk of mechanical damage to most artifacts and paintings. Class B is considered the reference for most of museums [2], since mechanical damage is proved to be avoided for RH values not exceeding the range of $50 \pm 15\%$.

The Lifetime Multiplier (LM) considers the risk of chemical degradation taking into account the activation energy of the degradation processes involved in the deterioration of the organic materials (i.e., 70 kJ/mol for yellowing of varnishes and 100 kJ/mol for degradation of cellulose). This index is a multiplier of the time left to an object to remain usable when compared to standard conditions of T = 20 °C and RH = 50%. Since the instantaneous values of LM exponentially depend on temperature (Equation (1)), the influence on chemical degradation of T variations is greater than that exerted by RH variations of the same magnitude [36]:

$$LM_i = \left(\frac{50\%}{RH_i}\right)^{1.3} \cdot e^{\left(\frac{E_a}{R}\left(\frac{1}{T_i} - \frac{1}{293.15}\right)\right)} \tag{1}$$

where RH_i is the instantaneous measured value of relative humidity at time i, T_i is the instantaneous measured value of temperature (expressed in K) at time i, E_a is the activation energy for the degradation of paper (100 kJ $\cdot$ mol^{-1}) and R is the perfect gas constant (8.314 J $\cdot$ mol^{-1} $\cdot$ K^{-1}). The level of risk associated to the Lifetime Multiplier values can be defined as follows [18]: safe when LM > 1, medium risk when 0.75 < LM ≤ 1 and high risk if LM ≤ 0.75.

2.4. The Hygrothermal Conditions Expected Indoors in the Period 2041–2070

As a consequence of the climate change scenario, the southern European regions will probably increase their need for summer cooling (while decreasing winter heating) in order to keep the environmental conditions suitable for artwork conservation [26]. To evaluate the effects of the climate change scenario in the residential building near Valencia and on the effectiveness of the microclimate frame, we followed the approach applied in [23,24] to forecast the expected indoor T and RH levels.

The principal steps of the methodology can be summarized as follows:

1. monitoring of the simultaneous indoor (a) and outdoor (b) climate over at least one year;
2. derivation of the outdoor/indoor heat and moisture transfer functions (TFs) through the building;
3. extraction of the outdoor climate in the interested area from a simulated scenario;
4. inverse modeling of the future indoor climate based on the derived TFs; and
5. evaluation of the expected changes for artwork conservation by means of damage functions.

The annual hygrothermal data monitored in the residential building (Step 1a) and outdoor (Step 1b) were used to derive the seasonal cycles of temperature and mixing ratio of moist air. The observations collected in 2016 during the heating period were discarded in the analysis in order to consider only the environmental conditions not affected by the HVAC systems.

The annual cycles of temperature and mixing ratio were fitted as generic time-dependent sinusoidal equations as follows:

$$x(t) = \bar{x} + \Delta x \cdot sin(\omega t - \Phi) \tag{2}$$

where x is the variable considered (i.e., T or MR), t is time (in days), $\bar{x}$ is the annual average of x, ω is the angular frequency (i.e., $\omega = 2\pi/P$ where P is the period, equal to 365 days) and Δx and Φ are the amplitude and the phase shift of the best-fit sine function, respectively.

The measured indoor and outdoor data were used to fit the annual cycles (Equation (2)) that regulate heat and moisture exchanges across the building envelope on a seasonal basis, obtaining the indoor coefficients, Δx_{in} and Φ_{in}, and the outdoor ones, Δx_{out} and Φ_{out}. The combination of the two sinusoids, i.e., the outdoor T or MR cycles in abscissa and the indoor T or MR cycles in ordinate, gives the annual hysteresis cycle in the building [23]. During the annual cycle, the capability of the building to accumulate or release heat and moisture is an important factor that influences the transfer functions (TFs) and can be expressed in terms of the gain of the building (A_B), defined as the ratio between Δx_{in} and Δx_{out}, and the phase shift (Φ_B), defined as the difference between Φ_{in} and Φ_{out} (Step 2).

Temperature and relative humidity daily data in the area of Valencia for the 30-year time window from 2041 to 2070 (Step 3) were extracted from the ENSEMBLES dataset [28]. The ENSEMBLES simulation model was developed within the ENSEMBLES European project (2004–2009) [37] to produce regional dynamic projections. The high-resolution projections used in this study were generated by the Max Plank Institute for Meteorology using the IPCC emission Scenario A1B [35]. Scenario A1B was developed by the Intergovernmental Panel on Climate Change (IPCC) and was chosen as it is a moderate scenario that assumes higher CO_2 emissions until 2050 and their decrease afterwards. The ENSEMBLES data were used to obtain the fitting coefficients Δx_E and Φ_E from the annual cycles (Equation (2)). To evaluate the effects of the outdoor climate scenario inside the residential building, the future hygrothermal conditions indoors were inversely simulated using derived T and MR transfer functions (Step 4) based on the same sinusoidal equations in Equation (2) with Δx and the Φ calculated using the gain A_B and the phase shift Φ_B of the residential building computed as described in Equations (3) and (4) and with indoor annual mean $\bar{x}$ estimated as described in Equation (5):

$$\Delta x = A_B \cdot \Delta x_E \tag{3}$$

$$\Phi = \Phi_B + \Phi_E \tag{4}$$

$$\bar{x} = \frac{\bar{x}_{in}}{\bar{x}_{out}} \cdot \bar{x}_E = \bar{x}_B \cdot \bar{x}_E \tag{5}$$

where A_B and Φ_B are the gain and the phase shift of the building, respectively; $\bar{x}_E$, Δx_E and Φ_E are the mean, the amplitude and the phase shift of the best-fit sine function calculated from the ENSEMBLES dataset (2041–2070), respectively.

The indoor RH values were computed after the simulated indoor T and MR data by applying the formula in [33]. The T and RH conditions expected in the residential building were finally used to determine the possible changes in the future risk of chemical deterioration for paper in terms of the Lifetime Multiplier index (Step 5).

3. Results and Discussion

The internal response of the microclimate frame to the external forcing of the room was explored by taking into account annual time series of observations of T and RH values collected in two different sites: the first series was registered in the Pio V Museum from 1 June 2014 to 31 May 2015 (hereafter called Museum) and the second from 15 February 2016 to 14 February 2017 in a residential building near Valencia (hereafter called Private).

The box-and-whiskers plots of the T observations (Figure 2a) show that the values inside the microclimate frame fully overlap the room ones in both the sites. The Wilcoxon–Mann–Whitney test was performed for each pair of microclimate frame (MF) and room (R) temperature and relative humidity series, both in Museum and in Private. The test assumes the samples are not normally distributed and the significance level was set to 5%. No significant difference was found between R and MF temperature series collected in the same site ($p > 0.05$); conversely, at both sites, the RH medians inside the microclimate frame significantly diverge from the room ones ($p < 0.0001$). The medians of the rooms are consistent and equal to 22.7 °C in Museum and 22.8 °C in Private; on the contrary, the variability associated to each dataset is significantly different, i.e less than ± 2 °C from the median in Museum and ranging from 16.5 °C to 31.0 °C in Private. While internal MF temperatures have the same variability as the external room, the internal RH levels are kept extremely stable throughout the year thanks to the buffering agent preconditioned to RH = 40% before being enclosed within the microclimate frame. The box-and-whiskers plots of RH values (Figure 2b) show a significant difference between the external (R) and internal (MF) distributions of data: in the rooms the range of the RH values registered is roughly between 20% and 60% with RH medians equals to 47%, while in both the sites the internal RH values are tightly kept around 40 ± 3% throughout the year. The few outliers found in the datasets (less than 1% of the total) were not discarded in the following analysis.

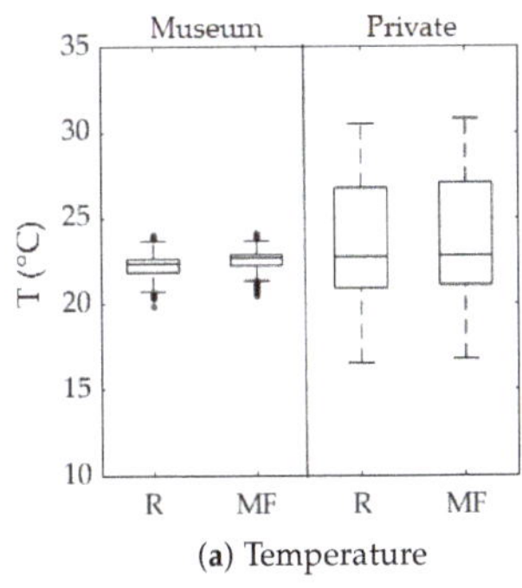
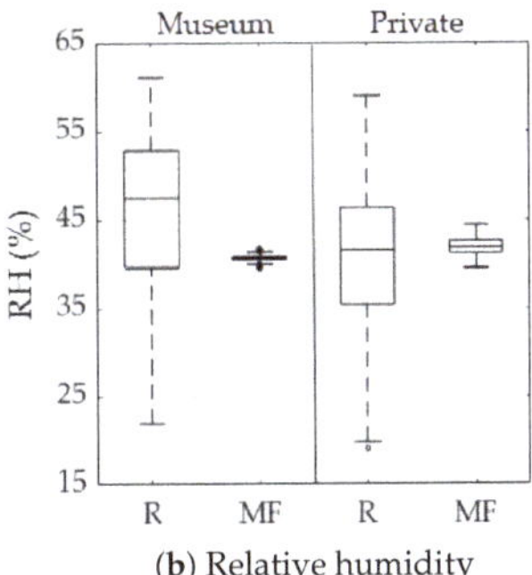

(a) Temperature   (b) Relative humidity

Figure 2. Box-and-whisker plot of temperature (**a**) and relative humidity (**b**) inside the microclimate frame (MF) and in the surrounding room (R) throughout a solar year. Outliers are indicated as points.

The hygrothermal response of the microclimate frame was further explored by comparing their internal conditions to the simultaneous room ones, as shown in the scatterplots in Figure 3. The two indoor environments are thermally controlled by active systems: in Museum, a HVAC system continuously controls temperature maintaining thermal stability with a minimal setpoint adjustment from winter to summer; in Private, an intermittent heating system is active only during the cold season, without any cooling in summer. Both in Museum and in Private, the internal MF temperatures closely follow the surrounding room conditions with a minor delay. In Private, when the heating system is switched off, the internal MF temperatures perfectly match those external, meaning that the thermal cycle is transferred unchanged inside the microclimate frame. The performance of the buffer is examined by relating the internal RH values to the external R temperatures: in Museum, where T is kept almost stable, the considerable RH variability of the room is tightly controlled inside the MF; in Private, as a consequence of the variability of the R temperatures, the internal RH values show a minor drift despite being below significance (Figure 3, lower panels).

Table 2 shows the results of the ASHRAE classification. Both Museum and Private rooms are associated to ASHRAE Class D, which protects only from mold growth with RH < 75% [17]. The employment of microclimate frame in Museum made it possible to reach Class AA, providing the best possible microclimate for the preventive conservation of the paintings (Table 2); in Private only Class B could be achieved, which however is considered the reference to prevent from mechanical damage [2] as it provides no risk for many artifacts and most books even if a moderate risk for high vulnerability artifacts and paintings remains. Analyzing the T and RH data collected in the Private, the amount of observations overcoming the tolerance bands of Class AA is significantly reduced, passing from 55% of the R values fitting into the required specifications to 90% of the MF ones.

Table 2. Attribution of the ASHRAE class of climate control in the four locations.

	ASHRAE Class	
Position	**Museum**	**Private**
Room	D	D
Microclimate frame	AA	B

In Figure 4, the bands of tolerance for ASHRAE Class AA are plotted together with the measured hourly data, better explaining the conditions established inside the microclimate frame in comparison to the surrounding room. Class AA considers a seasonal adjustment of ±2 °C respect to the annual mean with short-term fluctuations from the seasonal 91-day central moving average smaller than ±5 °C for temperature and no seasonal adjustment respect to the annual mean of relative humidity with short-term fluctuations below ±5%. Both in Museum and in Private, in the room environments, the observed RH fluctuations reach up to ±20%; however, simultaneous RH values inside the MF are kept reliably around 40%. In Private (Figure 4b), temperatures exceed 30 °C in summer and are occasionally below the lower tolerance band in winter. These values, being transferred inside the microclimate frame, are responsible for the impossibility to achieve ASHRAE Class AA as they are not compatible with conservation.

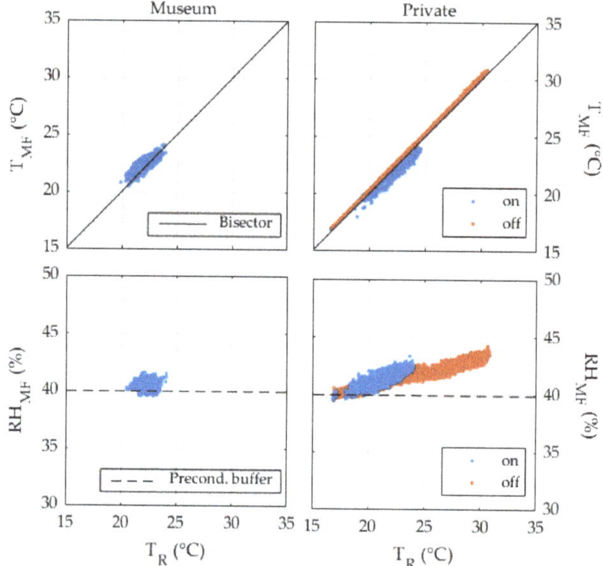

Figure 3. Scatter diagram of simultaneous temperature (T_R) and relative humidity values (RH_R) in the rooms versus values inside the microclimate frame (T_{MF} and RH_{MF}) during a solar year. In Private, the data points are grouped based on whether the intermittent heating system is active (on) or not (off).

The risk of chemical deterioration in the two sites was assessed through the Lifetime Multiplier, as shown in Figure 5. During winter, the LM values associated to both the rooms are higher as a consequence of the considerable drop in RH. However, it has to be highlighted that RH values below 30% may be dangerous for paper conservation (particularly when handling is foreseen) because at low moisture content the flexibility decreases while the brittleness increases [20]. In Museum, where temperature is controlled, an improvement in the duration expectancy of paper objects was observed inside the MF thanks to the buffering in RH values. In Private, the hot summer temperatures account for the almost unchanged average LM values obtained inside the devices. This result is justified by the the greater effect exerted by a drop in temperature on the increase in the life expectancy of an object with respect to the beneficial effect due to an equal drop in relative humidity [36].

The heat and moisture sinusoidal transfer functions in Private were determined as described in Section 2.4. The T and RH data monitored in the residential building were chosen as it is unconditioned for most of the year as well as being the conservation site when the study was conducted. The observations collected during the heating period were discarded in the following analysis. Two sinusoidal equations were fitted to the outdoor data in the area of Valencia [35] and the indoor values monitored in the room. Figure 6 shows the combination of the outdoor T and MR data (i.e., T_{out} and MR_{out}) in abscissa, with the indoor ones (i.e., T_{in} and MR_{in}) in ordinate. The coupled indoor and outdoor sinusoidal fits form the hysteresis cycle during the year. For temperature, it has the shape of an ellipse due to the thermal inertia of the building envelope and the building use and the T phase shift ($\Phi_{B,T}$) is 0.27. For mixing ratio, the yearly cycle is a straight line, meaning that the indoor MR conditions reach a rapid equilibrium with the outdoor ones; indeed, the MR phase shift is

$\Phi_{B,MR} = 0.01$, equal to a delay of about half a day for moisture transfer. It is worth noticing that the results could have been partially affected by the derivation of the transfer functions from the reduced dataset (i.e., including only the period not affected by the heating).

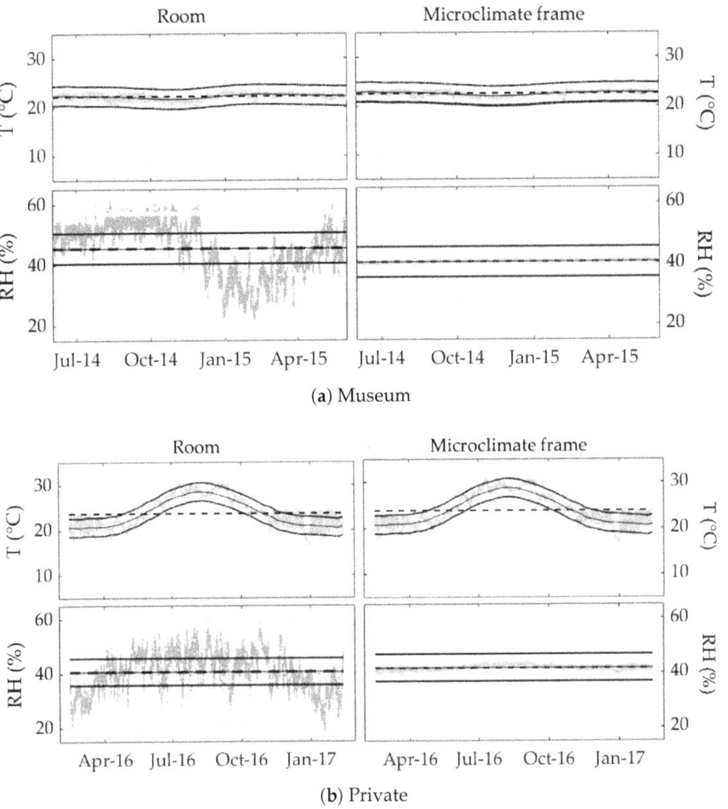

Figure 4. Temperature and relative humidity the bands of tolerance for ASHRAE Class AA (thick lines) together with the values measured in Museum (**a**) and Private (**b**) during a year (grey points). The thin lines indicate the seasonal moving average, the dashed lines the annual mean of the measured values.

The ENSEMBLES scenario for the period 2041–2070 in the area of Valencia forecasts an increase in the outdoor temperature of about +3.5 °C and an increase in the outdoor mixing ratio of about +0.7 g/kg, resulting in an average decrease of 7% in the outdoor RH. The heat and moisture transfer functions allowed simulating the indoor T and MR conditions inside the residential building: in the same 30-year window, the annual average levels are expected to be 27.5 °C for temperature and 37.5% for relative humidity. The potential chemical risk associated to these hygrothermal conditions was assessed through the Lifetime Multiplier values. As shown in Figure 7, the expected change in the indoor climate would lead to augmented chemical risk for cellulose during spring and to improved environmental quality of conservation in autumn. In the hypothesis of the maintenance of the use of the passive microclimate frame with a stable internal RH around 40%, the expected thermal level within the MF would keep the risk of chemical deterioration constant at LM = 1.3, meaning an extended lifetime expectancy for the painting if compared to the standard conditions (T = 20 °C, RH = 50%).

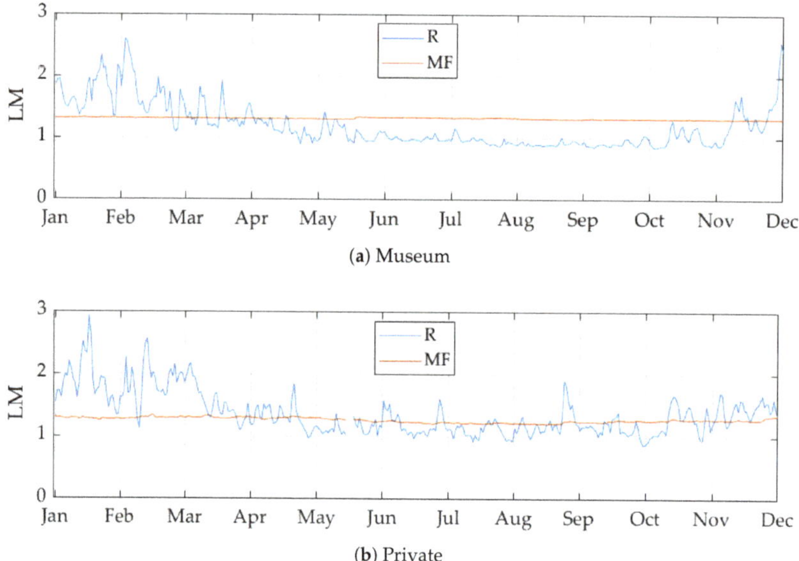

(**a**) Museum

(**b**) Private

Figure 5. Lifetime Multiplier values (LM) associated to the hygrothermal conditions over a solar year in Museum (**a**) and Private (**b**), in the room (R, blue) and inside the microclimate frame (MF, orange).

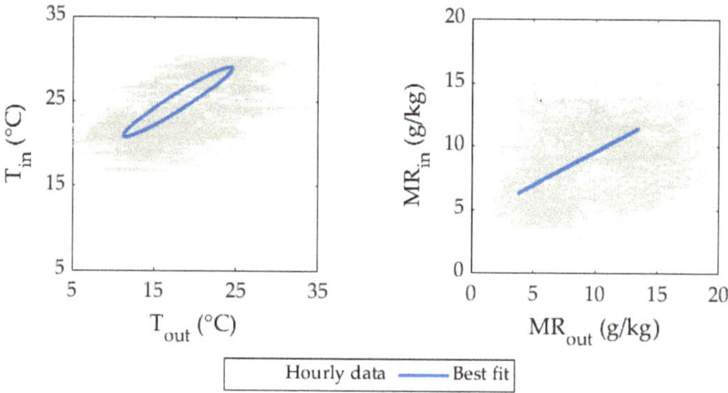

Figure 6. Indoor versus outdoor temperature (**left**) and mixing ratio (**right**) daily data over the solar year monitored (grey dots). The best fit lines (in blue) describe the yearly cycle inside the building.

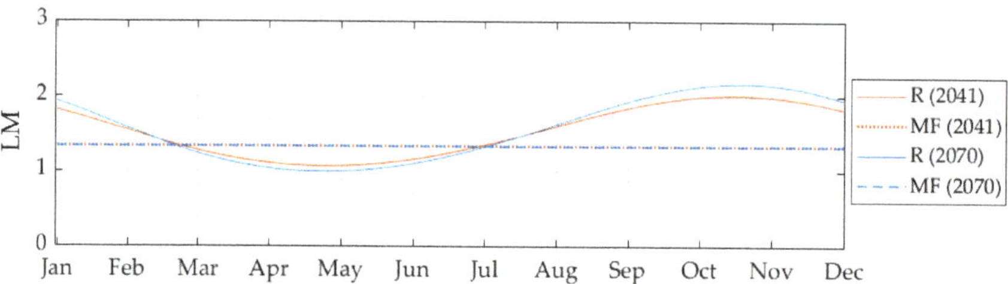

Figure 7. Lifetime Multiplier (LM) associated to the expected climate in Private in 2041 (orange) and 2070 (blue) in the room (R, solid lines) and inside the microclimate frame (MF, dotted/dashed lines).

4. Conclusions

The hygrothermal response of a passive microclimate frame hosting a portrait on cardboard by Sorolla was investigated by analyzing its internal T and RH conditions in response to the surrounding room environment. Its behavior was monitored in two different sites, i.e., the Pio V Museum of Valencia (Museum), with temperatures kept almost stable during the year, and a residential building (Private) in the same area, with a heating system active only in winter. The microclimate frame (MF) proved to be highly effective in controlling the internal RH levels but to be strongly influenced by the boundary thermal variability of the room. The ASHRAE classification of climate quality for conservation pointed out that the hygrothermal conditions in both Museum and Private would have prevented the painting only from the risk of dampness (Class D). On the contrary, within the microclimate frame, since the most dangerous seasonal RH cycles and short-term RH fluctuations were filtered out, the internal MF conditions were found to be compatible with ASHRAE Class B in Private and with ASHRAE Class AA in Museum, ensuring the best possible protection for the artifact. Moreover, the risk of chemical degradation for cellulose was assessed through the Lifetime Multiplier index, which confirmed that the microclimate frame is capable of better mitigating the risks in environments where temperature levels are adequate for conservation (Museum).

To extend our analysis to the application of the microclimate frame in the future, this study showed an example of how the indoor climate can be simulated in unconditioned buildings. In view of the climate ENSEMBLES-A1B scenario for the period 2041–2070 in the area of Valencia, this approach provided insight of the future hygrothermal conditions in Private. Even if the outdoor climate scenario is likely to be beneficial to the conservation of paper indoors in autumn, an increased risk of cellulose degradation would probably be observed during spring. The adoption of passive microclimate frames in the future expected conditions indoors would thus be an effective preventive conservation measure but it is yet important to be aware that these passive enclosures may not be sufficient to fully avoid the chemical degradation risk if an additional mitigation of the unsuitable temperatures is not provided. Considerable improvements on the current and future indoor climate might be provided by implementing some beneficial practices in the management of the environment. For example, the windows' opening might be rescheduled in order to enhance natural ventilation and a cooling device might be helpful to reduce the summer temperature peaks. In addition, passive retrofit intervention on the building envelope may be considered in order to relieve the expected effects of the climate change scenario in a sustainable manner. Notwithstanding the fact that the microclimate frames do not affect the user experience in terms of their overall dimensions and appearance, the adoption of these devices should imply that the ordinary management is adjusted according to their specific features. Indeed, large temperature fluctuations in the surrounding space may cause the absorption/release of considerable amounts of moisture by the buffering agent, determining the possibility of moisture exchanges with the painting itself. This means that it is fundamental to be aware of the effect over moisture exchanges exerted by temperature, which is not controlled within these passive enclosures. Moreover, since buffers are susceptible to ageing and loss of their buffering properties, it is fundamental to recondition and/or replace them on a regular basis in order to preserve their effectiveness.

Author Contributions: All authors helped to develop the manuscript. Conceptualization, Formal analysis and Writing—original draft, E.V. and F.F.; Supervision, Writing—review and editing, A.M.S.; and Resources and Data curation F.-J.G.-D.

Funding: This project received funding from the European Union's Horizon 2020 research and innovation programme under grant agreement No. 814624. This research was partially supported by the Plan Nacionalde I+D, Comisión Interministerial de Ciencia y Tecnología (FEDER-CICYT) under project HAR2013-47895-C2-1-P.

Acknowledgments: The authors thank CollectionCare for the funds for the publication, the Museum Pio V of Valencia, the staff of the IVACOR and the owners of the paintings. E.V. thanks Sapienza Università di Roma for the internship mobility grant.

Conflicts of Interest: The authors declare no conflict of interest. The funders had no role neither in the design of the study, in the collection, analyses, or interpretation of data, nor in the writing of the manuscript or in the decision to publish the results.

Abbreviations

The following abbreviations are used in this manuscript:

AEMET	National Agency of Meteorology of the Spanish Government
ASHRAE	American Society of Heating, Air-Conditioning and Refrigerating Engineers
CO_2	Carbon dioxide
HVAC	Heating, Ventilation and Air-Conditioning
IPCC	Intergovernmental Panel on Climate Change
IVACOR	Valencian Institute of Conservation and Restoration
LM	Lifetime Multiplier
MF	Microclimate Frame
MR	Mixing ratio of humid air (g/kg)
RH	Relative humidity (%)
T	Temperature (°C)
TF	Transfer Function
UPV	Polytechnic University of Valencia

References

1. Camuffo, D. *Microclimate for Cultural Heritage. Conservation, Restoration, and Maintenance of Indoor and Outdoor Monuments*, 2nd ed.; Elsevier: Amsterdam, The Netherlands, 2014; ISBN 9780444632968.
2. Bratasz, Ł. Allowable microclimatic variations in museums and historic buildings: Reviewing the guidelines. In *Climate for Collections: Standards and Uncertainties*; Ashley-Smith, J., Burmester A., Eibl, M., Eds.; Doerner Institut: Munich, Germany, 2013; pp. 11–19.
3. Luciani, A. Evolution of thermo-hygrometric standards. In *Indoor Environment and Preservation Climate Control in Museums and Historic Buildings*; Nardini: Florence, Italy, 2011; ISBN 9788840443393.
4. EN 15999-1:2014. *Conservation of Cultural Heritage - Guidelines for Design of Showcases for Exhibition and Preservation of Objects—Part 1: General Requirements*; European Committee for Standardization: Brussels, Belgium, 2014.
5. Bickersteth, J. IIC and ICOM-CC 2014 declaration on environmental guidelines. *Stud. Conserv.* **2016**, *61*, 12–17. [CrossRef]
6. Perino, M. Air tightness and RH control in museum showcases: Concepts and testing procedures. *J. Cult. Herit.* **2018**, *34*, 277–290. [CrossRef]
7. Shiner, J. Trends in microclimate control of museum display cases. In Proceedings of the Museum Microclimates: Contributions to the Copenhagen Conference, Copenhagen, Denmark, 19–23 November 2007; pp. 19–23.
8. Camuffo, D.; Sturaro, G.; Valentino, A. Showcases: A really effective mean for protecting artworks? *Thermochim. Acta* **2000**, *365*, 65–77. [CrossRef]
9. Bernardi, A.; Becherini, F.; Romero-Sanchez, M.D.; Lopez-Buendia, A.; Vivarelli, A.; Pockelé, L.; De Grandi, S. Evaluation of the effect of phase change materials technology on the thermal stability of Cultural Heritage objects. *J. Cult. Herit.* **2014**, *15*, 470–478. [CrossRef]
10. Dahalin, E. (Ed.) Improved Protection of Paintings during Exhibition, Storage and Transit. In *PROPAINT-Final Activity Report*; Norwegian Institute for Air Research: Kjeller, Norway, 2010.
11. Richard, M. Further Studies on the Benefits of Adding Silica Gel to Microclimate Packages for Panel Paintings. In *Facing the Challenges of Panel Painting Conservation: Trends, Treatments, and Training, Proceedings of the a Symposium at the Getty Centre, Phenix, Los Angeles, CA, USA, 17–18 May 2009*; Getty Conservation: Los Angeles, CA, USA, 2011.
12. Ferreira, C.; de Freitas, V.P.; Ramos, N.M.M. Influence of hygroscopic materials in the stabilization of relative humidity inside museum display cases. *Energy Procedia* **2015**, *78*, 1275–1280. [CrossRef]

13. Thickett, D.; Fletcher, P.; Calver, A.; Lambarth, S. The effect of air tightness on RH buffering and control. In Proceedings of the Museum Microclimates: Contributions to the Copenhagen Conference, Copenhagen, Denmark, 19–23 November 2007; pp. 245–251.

14. Sozzani, L. An economical design for a microclimate vitrine for paintings using the picture frame as the primary housing. *J. Am. Inst. Conserv.* **1997**, *36*, 95–107. [CrossRef]

15. Michalski, S. The ideal climate, risk management, the ASHRAE chapter, proofed fluctuations, and towards a full risk analysis model. In *Experts Roundtable on Sustainable Climate Management Strategies*; Getty Conservation: Los Angeles, CA, USA, 2007; pp. 1–19.

16. EN 15757:2010. *Conservation of Cultural Property—Specifications for Temperature and Relative Humidity to Limit Climate-Induced Mechanical Damage in Organic Hygroscopic Materials*; European Committee for Standardization: Brussels, Belgium, 2010.

17. American Society of Heating, Refrigerating and Air-Conditioning Engineers (ASHRAE). *ASHRAE Handbook—HVAC Applications: Chapter 24—Museums, Galleries, Archives and Libraries*; ASHRAE: Atlanta, GA, USA, 2019.

18. Martens, M. Climate Risk Assessment in Museums: Degradation Risks Determined from Temperature and Relative Humidity Data. Ph.D. Thesis, Eindhoven University of Technology, Eindhoven, The Netherlands, 2012.

19. Huijbregts, Z.; Kramer, R.P.; Martens, M.H.J.; Van Schijndel, A.W.M.; Schellen, H.L. A proposed method to assess the damage risk of future climate change to museum objects in historic buildings. *Build. Environ.* **2012**, *55*, 43–56. [CrossRef]

20. Menart, E.; De Bruin, G.; Strlič, M. Dose–response functions for historic paper. *Polym. Degrad. Stab.* **2011**, *96*, 2029–2039. [CrossRef]

21. Kompatscher, K.; Kramer, R.P.; Ankersmit, B.; Schellen, H.L. Intermittent conditioning of library archives: Microclimate analysis and energy impact. *Build. Environ.* **2018**, *147*, 50–66. [CrossRef]

22. Rajčić, V.; Skender, A.; Damjanović, D. An innovative methodology of assessing the climate change impact on cultural heritage. *Int. J. Archit. Herit.* **2018**, *12*, 21–35. [CrossRef]

23. Bonazzi, A.; Merlo, C.; Campana, F.; Bertolin, C.; Camuffo, D. Past, present and future effects of climate change on a wooden inlay bookcase cabinet: A new methodology inspired by the novel European Standard EN 15757:2010. *J. Cult. Herit.* **2013**, *15*, 26–35. [CrossRef]

24. Bertolin, C.; Camuffo, D.; Bighignoli, I. Past reconstruction and future forecast of domains of indoor relative humidity fluctuations calculated according to EN 15757: 2010. *Energy Build.* **2015**, *102*, 197–206. [CrossRef]

25. CORDIS. CLIMATE FOR CULTURE—Damage Risk Assessment, Economic Impact and Mitigation Strategies for Sustainable Preservation of Cultural Heritage in the Times of Climate Change. Available online: https://cordis.europa.eu/project/rcn/92906/factsheet/en (accessed on 28 June 2019).

26. Leissner, J.; Kilian, R.; Kotova, L.; Jacob, D.; Mikolajewicz, U.; Broström, T.; Ashley-Smith, J.; Schellen, H.L.; Martens, M.; van Schijndel, J.; et al. Climate for Culture: Assessing the impact of climate change on the future indoor climate in historic buildings using simulations. *Herit. Sci.* **2015**, *3*, 38. [CrossRef]

27. García-Diego, F.J.; Verticchio, E.; Beltrán, P.; Siani, A.M. Assessment of the minimum sampling frequency to avoid measurement redundancy in microclimate field surveys in museum buildings. *Sensors* **2016**, *16*, 1291. [CrossRef] [PubMed]

28. AEMET. Climate Projections for the XXI Century: Dynamic Regional Projections Based on the MPI-REMO Model Using the IPCC Emission Scenario A1B. Available online: http://www.aemet.es/es/serviciosclimaticos/cambio$_$climat/datos$_$diarios (accessed on 28 June 2019).

29. Roldán, C.; Juanes, D.; Ferrazza, L.; Carballo, J. Characterization of Sorolla's gouache pigments by means of spectroscopic techniques. *Radiat. Phys. Chem.* **2016**, *119*, 253–263. [CrossRef]

30. IVACOR. Institut Valenciá de Conservació i Restauració de Béns. Dos dibujos de Joaquin Sorolla de la Familia Traver. 2014. Available online: http://www.ivcr.es/media/descargas/monografia-sorolla-familia-traver-w.pdf (accessed on 28 June 2019).

31. Diego, F.J.; Esteban, B.; Merello, P. Design of a hybrid (wired/wireless) acquisition data system for monitoring of cultural heritage physical parameters in smart cities. *Sensors* **2015**, *15*, 7246–7266. [CrossRef] [PubMed]

32. EN 15758:2010. *Conservation of Cultural Property—Procedures and Instruments for Measuring Temperatures of the Air and the Surface of Objects*; European Committee for Standardization: Brussels, Belgium, 2010.

33. EN 16242:2012. *Conservation of Cultural Property – Procedures and Instruments for Measuring Humidity in the Air and Moisture Exchanges Between Air and Cultural Property*; European Committee for Standardization: Brussels, Belgium, 2012.
34. ASTME 104-02. *Standard Practice for Maintaining Constant Relative Humidity by Means of Aqueous Solutions*; ASTM International: West Conshohocken, PA, USA, 2012.
35. Universitat Politècnica de València (UPV). Historical weather observations in Valencia. Available online: http://dataupv.webs.upv.es/datos-historicos-de-la-observacion-meteorologica-en-valencia/ (accessed on 28 June 2019).
36. Michalski, S. Double the life for each five-degree drop, more than double the life for each halving of relative humidity. In Proceedings of the Preprints of the ICOM-CC 13th Triennial Meeting, Rio de Janeiro, Brazil, 22–27 September 2002; James and James (Science Publishers) Ltd.: London, UK, 2002; pp. 66–72.
37. CORDIS. ENSEMBLES-Based Predictions of Climate Changes and Their Impacts. Available online: https://cordis.europa.eu/project/rcn/74001/factsheet/en (accessed on 28 June 2019).

MDPI

St. Alban-Anlage 66

4052 Basel

Switzerland

Tel. +41 61 683 77 34

Fax +41 61 302 89 18

www.mdpi.com

Climate Editorial Office

E-mail: climate@mdpi.com

www.mdpi.com/journal/climate